Leaders on every food-growing continent warn of conflict, mass migration and war over **food, water** and **climate change.**

The UNIPCC warned that a just half a degree increase in temperatures would lead to more than $54 trillion in damages.[1] Climate chaos damage in the US for 2018 exceeds $300 billion.

Reducing the Western diet's adverse impact on human health and the environment is one of the greatest challenges facing humanity.
– **David Tilman, Regents' Professor, University of Minnesota**

Insect biomass loss has plunged 80% in 30 years. Insect loss threatens the survival of mankind.[2] Unless we change our ways of producing food, insects will go down the path of extinction in a few decades.
– **Francisco Sánchez-Bayo** *Biological Conservation*, April 2019.

Food is the strongest lever to optimize human health and environmental sustainability on Earth. We face an immense challenge – to provide our world population with healthy diets from sustainable food systems. The outcome is dire. *– EAT-Lancet Commission, 2019*

Modern food production drives climate instability and ecosystem destruction. A radical transformation of the global food system is urgently needed. Failing immediate action, today's children will inherit a planet that has been severely degraded and where much of the population will increasingly suffer from malnutrition and preventable disease.[3] **Prof Walter Willet, MD, Harvard** *– EAT-Lancet Commission*

Food production is a major source of greenhouse-gas, the main user of fresh water, and a leading driver of biodiversity loss and land-use change. At the same time, unhealthy diets are behind growing rates of diabetes, heart disease and cancer. This points to a need to change farming and eating if the world is to sustainably feed 10bn people by 2050.[4]
– The Economist, 2019

Emerald Renaissance:

World Hunger Solutions
Healthier for People, Producers and our Planet

Mark R. Edwards

Renaissance era world, air and water

Dedication

To Profs. Walter Willett and Johan Rockström and their international team of 30+ superb health, and science and sustainability professionals who had the insight and courage to publish *Food in the Anthropocene: The EAT–Lancet Commission on Healthy Diets from Sustainable Food Systems.*

Thank you for so eloquently laying out the Global Grand Challenge for which Emerald Renaissance is designed.

And to my fabulous life-partner, Ann Ewen, who creates an Emerald Glow for all of us lucky enough to know her.

ISBN-13: 9781797801810

Science / Biotechnology / Food / Social Justice

Copyright © 2019 by Mark R. Edwards

Emerald Renaissance may be used for educational purposes.

Chapters

Fossil Foods Violence Drives Climate and Changes Colors

Fossil foods attack blue air, turning skies grey as black particulates soar.
Deep mines and pumps rip out natural resources, depleting global stores.
Meat and dairy animals poop, pee, burp and fart wicked carbon forms.
Farmers store, dump, bury and burn wastes that escape with each storm.

Walrus fry while polar bears die as sunshine melts white ice to blue sea,
Solar energy once reflected, now amplifies heat, driving poles ice-free.
Penguins and puffins flee but nowhere to go; nests gone, ice melting.
Winds aloft change jet streams; Europe, America and Asia sweltering.

Intense heat wilts crops, soils crack, blue turns brown as waterways dry.
Drought, thirst and pestilence destroy crops, intense sun and hot skies.
Families left homeless as vicious wildfires and storms expand in size.
Weather, hunger and fierce wars force climate migration as oceans rise.

Hot winters force forest green to brown, then to black by bark beetles.
Warm winters enable pest explosions as cotton fields weep with weevils.
High winter lows disrupt dormancy sleep, then drops fruit from trees.
Powerful pesticides poison indiscriminately, massacring birds and bees.

POTUS ends pollution protections and turns red lights to green.
Polluters dump, farmers waste, coal fires burn surreptitiously unseen.
Societies face famine if production continues its malicious devour.
What will our children do for food? We are close to the witching hour.

Our children deserve sustainable food production that is fossil free.
Not just climate independent, but climate positive to stay rising seas.
Healthier foods that break obesity, diabetes and Western disease.
Superior nutrition that will people, animals and our planet please.

Our future generations deserve not polluted perversions of color,
But a beautiful Emerald Renaissance that restores pigments we favor.

Forward

The Emerald Renaissance explores 50 strategic disparities between two methods for producing the most vital stored energy for human life; food. One method cultivates tiny, rootless plants; the other giant field crops.

One method supplies over a centillion, (10^{600}) consumers daily as it has for over 3 billion years with the healthiest and most sustainable food on our planet. Nature biocycles nutrients so global ecosystems never run out of food. Nature extracts nothing and creates zero waste or pollution.

Intensive mechanical agriculture, (IMA) is addicted to fossil resources. In only 70 years, IMA has consumed over half of the non-renewable resources it depends on for food production. Fossil food production extracts and consumes the precious resources it needs. Fossil foods are unhealthy and have caused a catastrophic epidemic of obesity, diabetes, CDH, cancers and caused millions of premature deaths yearly.

Fossil food production requires over ten times the energy to manufacture staple crops than it delivers in calories. While this may seem incredibly short sighted, this story gets worse. Big Ag and the USDA advocated for and received billions in annual subsidies to produce corn ethanol that requires more energy to make than the biofuel yields. Burning food for fuel makes no economic sense, while it raises food prices for millions.

Environmental factors do not favor fossil farmers as prior practice has already consumed over half of all cropland – abandoned due to erosion, extraction, salt invasion and exhaustion. Water has become scarce and expensive. Costly fossil fuels drive up food prices and climate change.

The Emerald Renaissance reinvents farming with abundance methods and freedom foods, which provide healthier food for consumers, farmers, farm families and our ecosystems. Abundant agriculture does not compete with IMA for fossil natural resources because it uses none. Abundance methods cultivate freedom foods with superior nutrition while emitting zero pollution and restoring health to our ecosystems.

Our Food Crisis in Vital Health Metrics

Emerald Renaissance

Vision. Solve the *Global Grand Challenge*: Cultivate sustainable, biodiverse, and nutritious foods while preserving natural resources and restoring health to our environment.

The Emerald Renaissance applies abundance methods to cultivate freedom foods that optimize 4P health. Nature provides the engine for our survival. All we must do is listen, learn and lean into action.

- **People** – sustainable, nutritious, affordable and diverse foods that remediate micronutrient deficiencies and defend against disease. Freedom foods can end micronutrient deficiencies, obesity and diabetes. They provide disease defense and therapeutic solutions when maladies penetrate defenses. Freedom foods embrace social justice by making healthy and affordable food available to everyone.

- **Producers** – improve farmer life quality. Enhance economic, social and health for farmers, farm families, farm animals and rural communities. Reduce farmer health and safety risk.

- **Preservation** – save precious resources and restore flourishing ecosystems for our future generations. Eliminate massive soil loss and pollution with abundance methods that restore degraded cropland. Reverse IMA's enormous water consumption and pollution by producing superior foods that produce 20% more fresh water than they use in cultivation.

- **Planet** – repair wounds from slash and gash IMA. Restore biodiversity and ecosystem health and vitality. Reverse IMA's 40% GHG pollution. Cultivate healthier food with no emissions that capture, recycle and reuse millions of tons of carbon and other nutrients from the air, wastewater and other waste streams.

Freedom foods include all products made today from plants, animals and petroleum. Microcrops produce food, feed, fiber, fertilizer and plant-based meats. The same biosystems that cultivate food produce a multiproduct suite that cleans air and water and restores ecosystems.

Our Food Crisis in Vital Health Metrics

Challenge for You

Imagine that you had $1 billion to invest and mitigating climate chaos topped your priority list. Would you invest in the IMA fossil food linear model that systemically extracts, consumes, wastes and pollutes or abundance that capture carbon and cleans air, water and ecosystems?

Monetizing carbon represents the most effective way to remove carbon from the atmosphere and halt global warming. Abundance methods create a circular economy with two carbon revenue streams: direct carbon exchange payments for each ton of carbon captured, plus the sale of a wide array of multicrop bioproducts.

The *Emerald Renaissance* lays out the architecture for our food future but does not provide all the answers. The architecture remains in its infancy and needs your help. No company practices abundance methods completely today, but many production systems work towards these goals. I have worked directly with teams that use core abundance elements that avoid fossil resource consumption, waste and pollution. The strongest breakthroughs we have made are in human, animal and plant nutrition and eco-remediation and restoration.

I taught food marketing at ASU for 39 years and served as a consultant for many of the large agribusiness, energy and technology firms. My analysis of our food supply chain from field to fork led to extreme alarm for our future generations. My deep concern motivated 17 books in the *Green Algae Strategy Series* that focus on healthy and sustainable food and energy with microcrops. These books, listed in Appendix I, are available free in color PDF to students, teachers and thought leaders.

We need the engagement of our next generation to disrupt the wasteful linear fossil food step function. Teachers and thought leaders are the vital link among generations that can convey the Emerald Renaissance value proposition. Together, we can create a new circular model that produces healthier food for our future generations.

Please share your ideas. Mark R. Edwards, drmetrics@gmail.com

Why the Emerald Renaissance?

The **Green Revolution's mechanical agri-wars** pitted wealthy farmers with large land holdings against small farmers. Unsurprisingly, wealthy farmers won the Green Revolution, with devastating outcomes for the health of people, animals, natural resources and ecosystems. Scale economies favored the wealthy. They mercilessly squashed small farmers, then gobbled up their cropland and consumed public resources.

Wealthy winners gained billions. Large agribusinesses now control 90% of US food production. They control food and agriculture policy, while they reap billions in government subsidies. The top five meat and dairy companies create more GHG emissions than ExxonMobil, Shell or BP.[5] Big Ag pollutes without consequence.

The **Emerald Renaissance** offers the antithesis to revolution – a thoughtful and peaceful learning from nature's incredible ability for biological regeneration and renewal. Abundance methods offer a means to repair the considerable violence and damage mechanical agriculture has inflicted on people, animals, plants and our ecosystems.

Emerald's historical meaning

Many civilizations have endowed **emeralds** with robust meaning; valuing emeralds for nature's power for constant rebirth and renewal.

- **Indian mythology, Sanskrit,** *the green of growing things.*
- **Ancient China, Feng Shui,** *the energy of growth, new beginnings.*
- **Persia,** *"smaragdus," the earth's rhythm of rebirth and renewal.*
- **Ancient Egypt,** *the energy of nature and of eternal life.*
- **China,** *living* a good life, happily in harmony with nature.
- **Aztec and Incas,** *wealth, prosperity, rebirth and renewal.*
- **Islam,** *verdant, inspires happiness and assures love from Allah.*
- **Middle Ages clergy,** *nature's renewal and symbol of divine faith.*

The Emerald Renaissance aligns with the emerald's historical meanings.

1. Our Food Crisis in Vital Health Metrics

*The greatest wealth is health. – **Virgil***

The key question for disruptive change is: "Why do we need to change our food supply chain?"

Human societies are only one or two generations away from extinction unless we protect and preserve health for people, farmers, natural resources and our environment.

The most critical stored energy on our planet, food, currently faces a global state of emergency. Our shared peril becomes disturbingly clear with key health metrics.

The Green Revolution invented intensive mechanical agriculture, (IMA), completely dependent on fossil resources and heavy machines.

IMA practice spread like wildfire globally because farmers could produce higher yields with less labor, at least in the early years. The Green Revolution did fulfill its primary mission – produce low cost protein to feed hungry people after the World War II devastation.

Unfortunately, key assumptions underlying the IMA design contained fatal errors. The unintended consequences have diminished or destroyed health and vitality for billions of unfortunate victims, consumed massive fossil resources, and polluted our atmosphere, water and ecosystems. It is hard to imagine a more destructive food production system.

We Americans inflicted these horrific agonies on ourselves. The Green Revolution flew like a boomerang that came back and smacked us in the face. Americans spend twice as much per person on healthcare than people in other rich nations, yet we die younger and endure substantially more injury, illness, disabilities and premature death.[6]

IMA's ecologically destructive practices and processed fossil foods may have caused more damage to the health of people, natural resources and our environment than the combined effects of the US wars in Korea,

5

Vietnam, Persian Gulf, Iraq, Syria and Afghanistan. The body count from fossil food disorders systemically rises every year.

Please consider the top-line evidence in vital health metrics.

Human health – The calorie-dense but nutrient deficient fossil foods in the Western diet have catastrophically damaged human health.

1. Unhealthy diets now pose a greater risk to morbidity and mortality than **unsafe sex, alcohol, drug and tobacco use** combined.[7]

Note: In studying, teaching and working throughout the food supply chain for over 50 years, this is the most chilling metric for me. The *EAT-Lancet Commission on Food, Planet, Health* engaged 30+ world-class scientists assess the Western diet and fossil foods. Their incredibly harsh assessment of our fossil food system is breathtaking, and true.

2. Since the dawn of the Green Revolution, fossil foods have created a global obesity epidemic where 2.2 billion people are overweight with a tenfold increase in obese children.[8] Obesity and overweight rates for American adults have more than doubled to 72%.[9]

3. American children suffer from one of the highest overweight rates worldwide, 33%. The obesity prevalence among US children is 200% higher than among Western Europeans.[10]

4. Junk food comprised 86% of ad spending on black-targeted programming and 82% on Spanish-language television in 2017.[11]

5. Half of US adults over 50 suffer from a chronic health condition and over 1 in 4 suffer from two or more chronic conditions.[12] Many chronic health conditions are directly related to the fossil foods diet, which costs citizens over $3.3 trillion a year.[13]

6. Fossil foods have decreased the cost of food in the US to about 9% of the household budget, but the high-caloric, nutrient deficient foods have doubled healthcare costs to above 18%.[14]

7. Roughly 20% of deaths worldwide are attributable to the unhealthy Western diet – high in calories, carbohydrates, sugar and salt but nutrient deficient.[15]

8. Over 60% of the calories consumed globally come from only three low-nutrient foods; wheat, rice, and maize (corn).[16]

9. The CDC predicts that 1 in 3 children born after 2000 in the US will become overweight and diabetic, 45 million more children.[17]

10. Stephanie Seneff at MIT is a leading scientist on brain function predicts, "At today's rate of fetal exposure to pesticides and herbicides, by 2025 one in two children will be autistic."[18]

11. Over 44 million Americans, including 12 million children, suffered food insecurity in 2017.[19]

12. About 85% of Americans suffer from micronutrient deficiencies, often result in preterm births. Preterm births can cause birth defects, developmental disabilities, mental retardation, reduced immunity, blindness, poor learning and premature death.[20]

13. The 380,000 premature births in the US, 1 in 10 births, is higher than any other wealthy country and resembles sub-Saharan Africa. Preterm birth is the most frequent cause of infant death and disability and costs the healthcare system $26 billion a year.[21]

14. One in six, or about 15%, of US children aged 3 through 17 years have a one or more developmental disabilities.[22]

15. Human sperm counts have declined 59% over the last 40 years for men in North America, Europe and Australia.[23]

16. More people are malnourished today, over 3 billion, than at any time in world history.[24] One billion are severely undernourished.

17. Combustion emissions in the US account for about 200,000 premature deaths per year and over 2 million life-years lost. Agricultural emissions are the leading cause of premature air pollution deaths in the US.[25]

18. Pesticides costs more than $45 billion annually from health care costs and lost wages.[26] Pesticide exposure causes 2 million lost IQ points and another 7,500 intellectual disability cases annually.[27]

19. The overuse of antibiotics in animal production has accelerated the development of antibiotic-resistant bacteria, which cause 76 million cases of food-born illnesses a year and over 5,000 deaths.[28]

20. Toxic residues contaminate over 70% of the food on American tables.[29] Poisons contaminate drinks such as milk, beer and wine at even higher rates, 95%.[30]

21. The CDC conducted a study of 9,282 US citizens and found pesticides in 100% of those tested.[31]

22. Over 88% of the corn and 93% of soybeans grown in the US are GMO monocultures with unknown health consequences.[32]

23. Farmers in the US applied 1.8 million tons of glyphosate last year, enough to cover each cultivated acre in the US with 0.8 pounds of the poison cocktail.[33]

24. Exposure to agri-pesticides, along with other endocrine disruptors found in plastic, costs the US over $340 billion annually.[34]

25. Life expectancy in the US, declined in 2017.[35] The Western diet contributes to millions of premature disabilities and deaths.

If a foreign country inflicted the disease and disability on Americans caused by the Western diet of fossil food, politicians and citizens would call for war against that power. Incredibly, not only did we inflict IMA damage on ourselves, we exported the practice to our global neighbors.

Eco-insensitive IMA practices also have ravaged natural resource health.

Natural resource health – The IMA fossil foods model was built on violent mechanical methods that work like strip mining – systemically extracting and wasting non-renewable resources. Fossil foods consume and waste billions of tons of resources every season, driving food costs.

26. IMA food production consumes about 50% of the planet's land surface globally, and about the same amount in the US.[36]

27. Mechanical methods put farmers in grave danger. American farmers ranked #1 for workplace injuries and suffered among the highest fatality rates – 27 deaths a year per 100,000 workers.[37]

28. Over 300,000 Indian farmers have committed suicide since 1995 when they could not pay their debts.[38]

29. IMA farmers know that 11 prime aquifers are likely to go dry within 20 years.[39]

30. Corn fields degrade 97 million acres of US cropland and consume trillions of gallons of irrigation water; 5.6 cubic miles per year.[40]

31. IMA runoff carries so much pollutive fertilizer, agri-chemicals and poisons that 80% of waterways in some states are unsuitable for human recreation.

32. More fertile cropland has been abandoned due to erosion, salt invasion or exhaustion from IMA practices, 2 billion hectors, than is under cultivation today, 1.5 billion hectors.[41]

33. Species extinction occurs now at 1,000 natural levels due to human actions; largely from IMA monocultures and pollution.[42]

34. Nearly 80% of all insect biomass has been extinguished in 30 years. Insect loss will cause a "catastrophic collapse of natural ecosystems and threaten the survival of mankind."[43] The authors conclude that *the world must change the way it produces food.*

Environmental health – IMA practices degrade and destroy our air, water, croplands and fragile ecosystems.

35. Since the beginning of the Green Revolution, 75% of plant genetic diversity has been lost.[44] Most are now extinct.

36. IMA emits 40% total GHG to the atmosphere with food processing, packaging, transportation, storage and retail sales.[45]

37. The top five meat and dairy companies create more GHG emissions than ExxonMobil, Shell or BP.[46]

38. Ocean temperatures in 2018 reached their highest ever recorded. The second warmest was 2017, followed by 2015, 2016 and 2014. The heat increases in the world's oceans in was equivalent to 388 times China's electricity generation.[47]

39. Groundwater pumping adds more than 25% to sea level rise and is accelerating.[48]

40. Global warming melts polar ice sheets far faster than expected. Greenland ice is melting four times faster than 15 years ago. Antarctica is losing six times more ice mass annually now than 40 years ago.[49] Ice sheet melt can raise sea levels 10 feet this century.

41. Water pollution from IMA fertilizers has created >400 dead zones globally that are expanding at 10% a decade.[50]

42. Meat animals produce 130 times as much excrement as the entire human population and pollute air, water and ecosystems.[51]

43. IMA farmers apply 5.6 billion pounds of pesticides globally and 2.2 billion pounds in the US each year, which kills flora, fauna and all types of soil microbes, insects and aquatic life.[52]

44. In only 55 years, industrial fishermen have extracted 92% of the ocean's top predators, severely disrupting marine ecosystems.[53]

45. About half of the world's tropical forests have been cleared, largely for IMA crop production.[54]

46. Over 75% of the world's food is generated from only 12 plants and five animal species.[55] More than 90% of crop varieties have disappeared from farmers' fields.

47. Carbon emissions in the US increased by 3.4% in 2018, the second largest annual gain in more than two decades.

48. Climate change is a deafening, piercing smoke alarm going off in the kitchen, The UN Doomsday Report, UNIPCC.[56]

49. If we don't take action, the collapse of our civilizations and the extinction of much of the natural world is on the horizon, David Attenborough, BBC's Planet Earth.[57]

These vital metrics raise the question:

> *Has there ever been a more critical time to re-invent our food supply in order to reverse each of these consequences?*

After 70 years of IMA practice, the metrics make clear that continuing fossil food production will make health severely worse for people, natural resources and our environment. A food production system that consumes and then wastes and throws away the critical resources it needs for production cannot be modified to address its countless fatal flaws. Warren Buffett offers sage advice.

> *In a chronically leaking boat, energy devoted to changing vessels is more productive than energy devoted to patching leaks.*　　　　　　　*- Warren Buffett*

The Emerald Renaissance and the adoption of abundance methods to produce freedom foods can reverse each of these grim metrics. Restoring health to people, natural resources and our environment may take several decades, but the time to change vessels is – now.

The next chapters describe the IMA design flaws and critical false assumptions that have created our broken IMA food supply. The analysis provides context for the key metrics and begins the process of quantifying the appalling cost. Metrics also provide a baseline to quantify the value of reversing IMA damage with abundance practices.

For impatient readers who want solutions now, you may go directly to Chapter 6, which introduces Emerald Renaissance solutions.

2. Old, Old Questions – Fresh Solutions

The Emerald Renaissance reinvents food production with
novel biological solutions that address ancient challenges,

The Emerald Renaissance tackles over 50 fascinating quandaries. Many are as enduring as humans have practiced agriculture, 11,000 years. Failure to answer these questions led to errors in practice that repeatedly levied the ultimate cost – community starvation, pestilence and war.

Inability to solve these challenges led to the extinction of hundreds of ancient civilizations. Our failure to find solutions quickly will lead to the expiration of modern human societies. As the *EAT–Lancet Report* correctly observes, the food situation is grim. A global transformation of our intensive mechanical agriculture, (IMA) system is urgently needed.[58]

IMA applies huge diesel machinery to denude broad expanses of natural ecosystems, then consumes massive fossil resources in order to force nature to do the farmer's will. Mechanical methods systematically degrade soil with ploughs, rippers, compactors, fertilizers and pesticides until the soil is exhausted. Then, IMA farmers abandon exhausted cropland, leaving millions of acres wasted for future generations. More farmland has been abandoned in the last 60 years than is farmed today.[59]

Intensive mechanical agriculture applies heavy farm machinery and:

1. Carves out nature - scrapes the natural ecosystem "clean" to make way for monocultures.

2. Slashes and gashes - rips the topsoil deeply; prepares soil bed.

3. Dashes - disks the topsoil flat to prepare for planting.

4. Mashes – crushes the soil with monstrous heavy equipment.

5. Degrades – with herbicides, planting, cultivation, weeding.

6. Squashes - with irrigation and wet-ground compaction.

7. Kills - all microorganisms with tons of fertilizer, herbicides, fungicides and pesticides.

8. Erodes - topsoil, fertilizers and pesticides that migrate into waterways and rural ecosystems.

9. Extracts - huge stores of macro, micronutrients and humus with every crop, without replacement.

10. Produces – fossil foods that suffer from nutrient dilution and hidden hunger and transfer micronutrient deficiencies to our children, and adults.

Old, Old Questions – Fresh Solutions

Abundance biological solutions work in harmony with nature to cultivate healthier, truly sustainable freedom foods with microcrops grown with free energy – sunshine.

Abundance methods allow nature to work effectively and effortlessly.

1. Build a cultivation raceway and cultivate microcrops.

2. Energy – gravity and sunshine.

3. Zero – cropland, fresh water, fossil fuels, chemical fertilizers, pesticides or herbicides.

4. Restores ecosystems – air, water, and biodiversity; flora and fauna.

5. Harvest – daily, 320 days a year.

6. Freedom foods are healthier for people, producers and our planet.

Freedom foods can be made in any form, shape, taste or color.

Emerald Renaissance

The Emerald Renaissance proposes a transformation that addresses one meta-question, the *Global Grand Challenge*:

How do we cultivate sustainable, biodiverse, and nutritious food that restores health to people, producers and our planet?

Old, Old Questions – Fresh Solutions

The meta-solution requires solving a series of longstanding challenges. **People** deserve healthier, sustainable and affordable foods. How do we:

- Cultivate foods that are cleaner, more nutritious and tastier, yet prevent rather than cause disease?
- Address social justice; where all people have access to good, affordable food?
- End micronutrient deficiencies and the hundreds of associated severe disabilities, diseases and mental problems?
- Prevent millions of preterm deliveries – low birthweight babies, stunting, wasting, autism and other developmental disabilities?
- Grow healthier food locally in any geography, even in large cities?
- Prevent and treat those suffering from obesity and diabetes?
- Prevent and treat many Western diet diseases, including heart, lung, brain, neurological disorders and cancers?
- Transform the entire food supply chain to end extraction, waste and pollution, and instead restore environmental health?

Producers deserve work that is less physically demanding and offers a substantially lower risk of accidents, disabilities, death and suicide. Producer families and farm animals should live free of fear from diseases and disabilities inflicted by local air, water and ecological pollution and insidious agri-poisons. How do we:

- End the costly and wasteful single-use, linear agriculture model?
- Reduce producer risk from crop failure?
- Relieve growers from the high cost of seeds, water, fossil fuels, fertilizer and agri-chemicals?
- Eliminate the use of monocultures and the direct destruction of biodiversity – insects, birds, reptiles, amphibians and fish?
- Enable growers to produce higher value multiproduct crops?
- Preserve stores of natural resources for future generations.
- Cultivate successfully independent of weather and climate chaos?

Old, Old Questions – Fresh Solutions

- Eliminate the substantial risks from heavy machinery, fatigue, dust, sun exposure and black soot particulates?
- Eliminate the extreme risks from agri-pesticides and poisons on farmers, farm families, animals and rural communities.
- End the risks and controversies associated with GMO crops.

Production technologies may appear neutral for food production. They are not. Healthy, sustainable foods are best grown by which technology?

Two Technology Sets to Produce Food	
Intensive Mechanical Fossil foods	**Abundance Freedom foods**
Mechanical solutions	Biological solutions
Crops: giants with roots	Tiny rootless
Micronutrient deficient	High nutrient density
Bioactive compounds = zero.	Over 200 bioactives
Unavoidable hidden hunger	Packed with nutrients
Produce one crop a year	Produce a crop daily
Experience: 70 years	3+ billion years
Abandoned cropland: >50%	Zero
Mass balance of -10:1	1:1
Massive waste	Zero
Emits tons of carbon	Captures C
Fossil energy dependent	Solar energy
GMO monoculture staples	Naturally biodiverse crops
Mature in a full season	Mature in a day, every day
Uses heavy equipment	Light electric gear
Enormous consumption	Zero, biocycles nutrients
Vast croplands	Zero
Massive irrigation	Zero
Tons of chemical fertilizer	Zero, biocycles nutrients
Tons of pesticides	Zero poisons
Decimates insects	Restores insect habitats
Massively pollutes	Repairs our environment
Destroys biodiversity	Restores flora and fauna

Old, Old Questions – Fresh Solutions

IMA produces fossil foods that impose horrific violence to the health of people, producers and our planet. Abundance methods cultivate freedom foods and serve as a steward for benevolent restoration and renewal.

Our Planet presents ugly deep scars from IMA. How do we:

- End strip mining cropland that degrades and destroys topsoil?
- Mend the gruesome scars and restore environmental health?
- Capture, recycle and reuse carbon and other nutrients?
- Terminate systemic nutrient extraction, waste and pollution?
- Eliminate fertilizer and pesticide erosion and toxin migration?
- Abolish massive animal GHG emissions and manure effluence?
- Bring dead, degraded and exhausted soil back to life?
- Restore rather than exterminate biodiverse flora and fauna?
- Regenerate habitat and food for birds, bees, frogs and fish?
- Revive and renew natural ecological systems.
- Flip the IMA negative agri-ecological footprint to net-positive?
- Produce renewable, biodegradable plastic, packaging and building materials that sequester rather than emit carbon?

Our **children and their children** deserve plentiful natural resource stores and restored clean, green ecosystems.

> *Abundance methods cultivate freedom foods and provide the engine for our survival and the assured vitality for our children, their children and many future generations.*

We can solve these challenges, together. We need a renaissance of thought to re-invent agriculture by listening to nature.

The *Emerald Renaissance* provides the first draft architecture with design elements that address each of these vital challenges. Please engage with your ideas to make our world better.

The next chapter provides a brief history of the Green Revolution.

3. The Green Revolution

A food system that systemically extracts, consumes and wastes the finite resources on which it depends for production can stand only until the first critical resource becomes extinct.

Key takeaways

- What is the Green Revolution linear production model?
- What are the design elements of Intensive Mechanical Agriculture?

The Green Revolution spawned Intensive Mechanical Agriculture, (IMA). IMA applied mechanical technologies against nature that extract massive fossil resources to resolve global hunger. After the horrific displacement and destruction created by the Second World War, leaders wanted to address global hunger.

The Green Revolution linear fossil foods model (below) did achieve its prime objective: produce sufficient protein to avoid mass starvation.

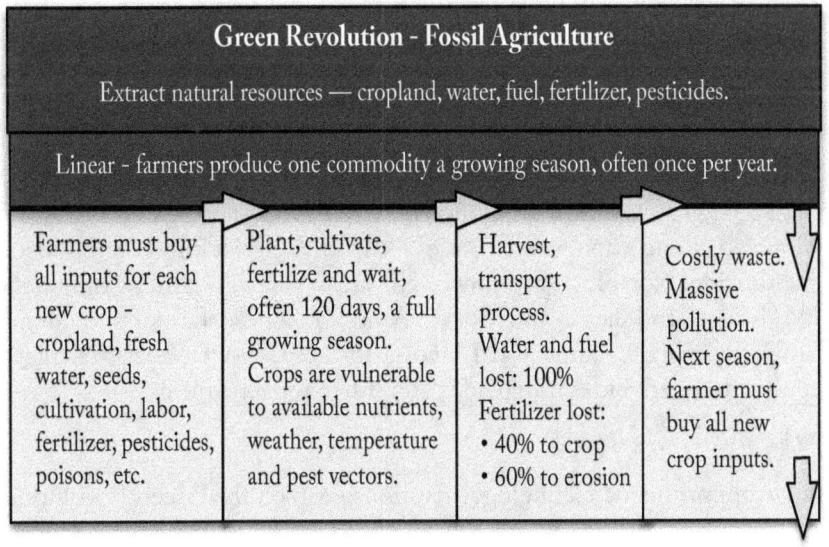

Green Revolution - Fossil Agriculture

Extract natural resources — cropland, water, fuel, fertilizer, pesticides.

Linear - farmers produce one commodity a growing season, often once per year.

| Farmers must buy all inputs for each new crop - cropland, fresh water, seeds, cultivation, labor, fertilizer, pesticides, poisons, etc. | Plant, cultivate, fertilize and wait, often 120 days, a full growing season. Crops are vulnerable to available nutrients, weather, temperature and pest vectors. | Harvest, transport, process. Water and fuel lost: 100% Fertilizer lost: • 40% to crop • 60% to erosion | Costly waste. Massive pollution. Next season, farmer must buy all new crop inputs. |

The Green Revolution

Fossil foods dominate production globally. The linear IMA food model extracts and consumes massive amounts of fossil resources, grows large-scale monocultures, uses chemical fertilizers and pesticides extensively, and produces meat in CAFOs (confined animal feeding operations).

Roughly 50% of the earth's land area is devoted to agriculture. Crops are planted on one-third while animals graze on two-thirds.[60] Forests are diminishing to IMA but occupy about 20% of the world's land area.

IMA uses primarily only 12 crops that overwhelmingly end up as animal feed, biofuels, or processed junk food ingredients.[61] The FAO reports that over 60% of the calories consumed by the world population come from only three nutrient-deficient foods; wheat, rice, and maize (corn).[62]

IMA uses mechanical equipment to violently strip cropland, much like strip mining. IMA monoculture methods steal from the earth but do not give back. IMA applies a linear design dependent on intensive fossil resource extraction and consumption to produce a single commodity crop, a staple, nutrient deficient food grain. The linear mode uses fossil resources inefficiently only once and discards the residuals to pollute air, water and cropland. New fossil agri-inputs must be repurchased every growing season. Resource use inefficiency and waste create enormous costs for farmers and extensive pollution and poisons to ecosystems.

The aftermath of WWII presented an opportunity to transfer the mechanical and chemical weapons of war to build intensive agriculture – munitions, heavy machinery and intensive use of fossil fuels.

The Green Revolution developed between 1950 and the late 1960s and doubled, and in some cases tripled, agricultural production worldwide. Notable exceptions were areas such as Africa, India, the Mid-East, Indonesia and South America. These farmers could not afford the higher annual IMA costs for seeds, water, fuels, fertilizer, pesticides, herbicides, fungicides and other agri-chemicals.

Technology applications

Five opportunistic technology adaptions enabled the Green Revolution.

The Green Revolution

1. **Transfer the weapons industry** into the agri-fertilizer and agri-chemicals industry to increase yields.

2. **Adapt the wartime heavy equipment** to agricultural production – land clearing, levelling, cultivation, weeding, fertilization, harvest and transportation to reduce human labor.

3. **Use fossil fuels intensively** to power heavy equipment, extract, mine, transport and apply inorganic fertilizer and agri-chemicals. Use more energy to process and store commodities and to transport food products to retail stores and then to consumers.

4. **Adapt military pumps and machinery** for moving water hundreds of miles for irrigation and for pumping irrigation water from deep underground aquifers. Intensive irrigation expanded cropland available for crop production. Irrigated lands are often four times more productive than lands that rely on rain because irrigation occurs in areas with more sunshine and water can be delivered precisely when the crop becomes thirsty.

5. **Develop high-yield hybrids of cereal grains**, especially dwarf wheat and rice. These grains took energy away from roots and stems and gave more energy to seeds, which are the only edible portion of the plant.

These technology adaptions worked together to improve the quantity and lower the cost of food staples.

Sustainable food system?

A sustainable food system would yield systematic increases in food productivity, reduce hunger and improve health. Unfortunately, the reverse is true because the five technologies that enabled the Green Revolution were opportunistic and not strategic.

These adapted technologies worked initially but food productivity improvements have plateaued or decreased. Each technology application was a one-time change. Each offer hope for only tiny incremental improvements in the future.

The Green Revolution

A recent study reports abrupt declines or plateaus in the rate of major crop production, which undermine optimistic projections of constantly increasing crop yields.[63] A third of total global food grains have experienced yield plateaus or abrupt decreases in yield, below.

The authors concluded: *"... we found widespread deceleration in the relative rate of increase of average yields of the major cereal crops during the 1990– 2010 period in countries with greatest production of these crops, and strong evidence of yield plateaus or an abrupt drop in rate of yield gain in 44% of the cases, which, together, account for 31% of total global rice, wheat and maize production."*

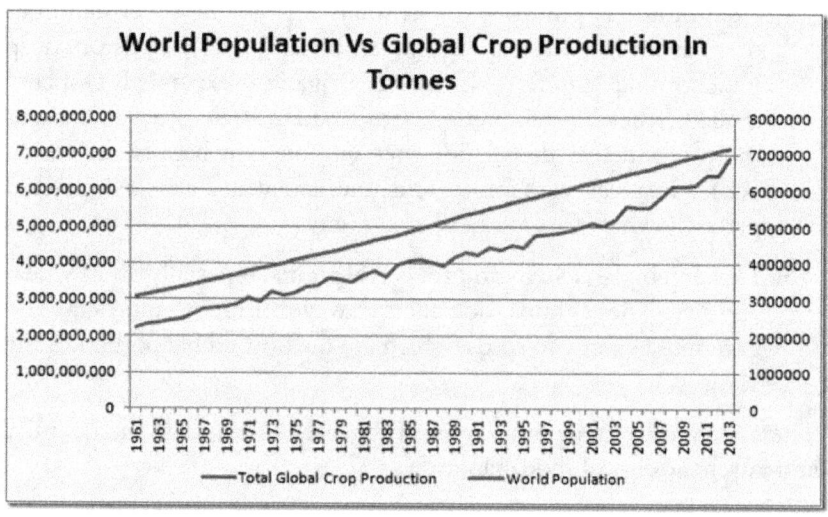

World population and crop production – Source: FAOSTAT

China provides a useful case study. Agricultural investment in China increased threefold from 1981 to 2000, but rates of increase for wheat yields have remained constant, decreased by 64% for maize (corn) and are negligible in rice. Similarly, the rate of maize yield has remained largely flat despite a 58% investment increase over the same period. Factors contributing to the plateaus or declines in food production rates include soil degradation, climate chaos, and the extensive use of fertilizers and pesticides.[64]

The Green Revolution

Optimists believe gene editing will renew food productivity increases, a hopeful strategy already called Green Revolution II.[65] Gene editing may allow increases in food production, but genetic engineering will ignite sharp debate over health, ethics, sustainability, and predictable unintended consequences. Genetics will do little for fossil resource consumption, waste and pollution. Genetics cannot overcome the severe problem of nutrient deficiencies in cropland, that transfer to fossil food.

World hunger solution?

The Green Revolution expanded food production with IMA, but today more people are hungry than ever before. Nearly three billion people are hungry and nearly a billion suffer from chronic undernourishment.[66]

IMA increased dependency on a few high-yield staples; white corn, soybeans, wheat and rice. These commodities are calorie-rich but nutritionally poor. The price of many foods fell, especially processed foods and sweetened beverages that could be made from corn or soy.

At the same time, the price of fresh foods such as fruit and vegetables increased substantially, in relative terms. Poor people made the obvious choice to buy cheap, highly processed foods that deliver calories but meager nutrition. These unfortunate consumers now suffer from obesity, diabetes, CHD, cancers and a litany of other Western diet diseases.

Hybrid and GMO crops increase food grains and are planted intensively

Crop cross-breeding, hybridization, for increased yield initially for increased yield and made farmers more money. Unfortunately, many important traits including flavor, color and nutritional value vanished.

The Green Revolution

The commodity crop focus had the unintended consequence that many of the nutritional advantages of diverse, fruit and vegetable-rich diets were lost.

Similar to yield increases in the early years of the Green Revolution, GMO benefits have diminished. GMO crops were approved with the promise that they would require fewer pesticides and herbicides because they had genetic enhancements that provided pest resistance. Improving yields for GMO crops weakened the organism. Shorter roots give plants less stability and the inability to reach deep for moisture and nutrients. Consequently, farmers must apply more irrigation and more fertilizer.

The many scientists that predicted weed resistance would be short lived were correct. Biological systems outsmarted Monsanto and the other GMO providers. Within only a decade, so many pests had developed resistance that now GMO crops require more pesticides than hybrids.[67]

The number of herbicide resistant weeds has increased over 45 years from zero to over 450. Increased weed resistance to GMO crops drives up the volume of herbicide needed each year by about 25%. The annual increase in herbicides required has grown from 1.5 million pounds in 1999 to 90 million pounds in 2011.[68] Those poisons migrate from the field on wind and water and kill indiscriminately in nearby ecosystems.

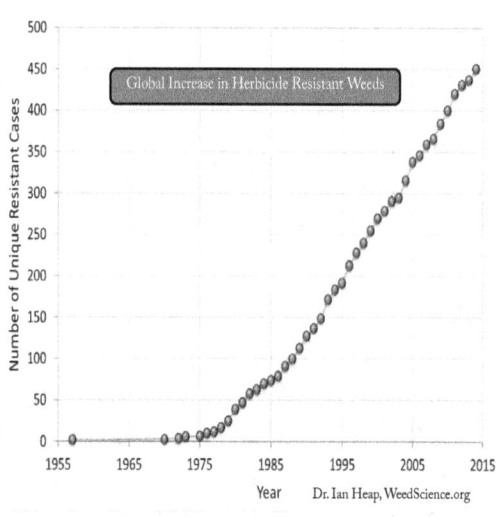

Dr. Ian Heap, WeedScience.org

Food processors made staple crops into vegetable oils, which were frequently hydrogenated to form trans-fats. The oils used in margarine, baked goods and high fructose corn syrup became ubiquitous to sweeten foods and drinks. Low-cost but health-challenging empty calorie choices increased with the availability of processed foods.

A recent study found that processed foods make up 58% of the calories consumed by Americans and contribute 90% of America's added sugar intake.[69] These foods include soft drinks, cereals, packaged baked goods, packaged snacks, confectionery and desserts, reconstituted meat products such as chicken or fish nuggets and instant noodles and soups

Currently, just 17 low-nutrient plant species are consumed as 90% of the global human diet.[70] IMA's dependence on just a few staple crops exported nutritional deficiencies globally through the Western diet. Consumers in both developing and developed countries, lack adequate vitamins, micro-nutrients and health-promoting phytonutrients.

Dependence on staples gave rise to hidden hunger caused by deficiencies in micronutrients, particularly iron and zinc, and essential vitamins, especially provitamin A.

Malnutrition expands

Today, 66% of the 7.6 billion world population is malnourished due to inadequate or unbalanced intake of nutrients. More people suffer from nutritional deficiencies than ever in human history.[71]

Currently, 99.7% of human food (calories) come from the land, while less than 0.3% comes from the marine and aquatic ecosystems.[72] Therefore, food production on land represents the primary opportunity to address world hunger.

Hunger and malnutrition have increased since 2012, returning to prevailing levels from a decade ago.[73] Nine out of ten countries face a serious burden from two or three forms of malnutrition; child stunting or wasting, anemia in women and obesity. The global cost for malnutrition exceeds $3.5 trillion per year.

Malnutrition takes 50% of child lives globally as well as extensive premature adult mortality.[74] Under nourishment creates life-time costs from impaired learning poor educational performance, weakened brains, hearts and lungs and increased health care costs.

The Green Revolution

Over 3 billion people suffer from lack of one or more key micronutrient. Every year and more than 3 million children die because they don't get enough food to eat.[75] About 9 million people globally die of hunger or malnutrition each year, while millions more perish unreported.

Malnutrition causes more illness than any other malady.[76] Children under five years face multiple malnutrition burdens: 155 million are stunted, 51 million are wasted, and 41 million are overweight. Over 20 million babies are born with abnormally low birth weight each year.[77]

Life-long consequences include diminished mental ability and learning capacity, severe fatigue, depression, poor school performance, reduced earnings and increased risks of chronic diseases.

Obesity

Calorie-dense but nutrient deficient modern fossil foods have created a global obesity epidemic. Overweight and obesity among adults are at record levels in 2015 with more than 2.2 billion people, 39% globally, are overweight or obese and suffer health problems because of their weight.

People affected by obesity has doubled since 1980. The US has the most obese adults, 80 million, followed by China, 57 million.[78]

The CDC reports that empty calories from added sugars and solid fats contribute to 40% of daily calories for US children and adolescents age 2–18 years—affecting the overall quality of their diets. Approximately half of these empty calories come from six sources: soda, fruit drinks, dairy desserts, grain desserts, pizza, and whole milk.

Minority children are at particular risk. Commercials for TV programming for black and Hispanic youth almost exclusively push fast food, sugary drinks, snacks and candy. Junk food comprised 86% of ad spending on black-targeted programming and 82% of spending on Spanish-language television in 2017.[79]

The number of obese children and adolescents globally has risen tenfold in the past four decades. If current trends continue, more children will be obese than moderately or severely underweight by 2022.

The Green Revolution

The Green Revolution created the obesity epidemic with cheap foods that provided energetic calories but meager nutrition. Poor families with limited incomes cannot afford healthy nutritious foods. Whole, unprocessed foods are too expensive.

> *People are fed by the food industry that pays no attention to health and treated by the medical industry that pays no attention to food.*
> *– Wendell Berry*

The prevalence of obesity in the US reached 40% and affected 93 million US adults in 2016.[80] Obesity-related conditions include heart disease, stroke, type 2 diabetes and certain types of cancer that are some of the leading causes of preventable, premature death. The estimated annual medical cost of obesity in the US was $147 billion. Obese people pay a $1,429 higher medical cost than those of normal weight.[81]

The cost of diabetes in 2012 was $245 billion, an increase of 41% over 2007.[82] The CDC predicts that 1 in 3 children born after 2000 will become diabetic, about 45 million more children.[83] Diabetes drag alone should motivate consumers to take action. But food suppliers continue to sell foods that are calorie dense, but nutrient poor.

Even in the US, the wealthiest nation on the planet, 20% of children live in food insecure households.[84] Over 31 million children live in impoverished households and enrolled in the National School Lunch Program.[85] More than 44 million Americans participate in the Supplemental Nutrition Assistance Program (SNAP) in order to acquire food products because they cannot afford retail food.[86]

Design flaw

The designers of the Green Revolution made a serious assumption error. They looked at rail-thin starving people and decided global hunger presented a protein crisis. Their fix may be called the "puff strategy." The designers decided to puff-up starving people with protein calories. Puff-up works short term to avoid starvation but a lack of micronutrients created many severe new medical issues.

The Green Revolution

Scientific analysis determined that providing a single nutrient, protein, was far from sufficient to address world hunger. Hungry people need protein **and** they also need the full set of essential nutrients to prosper.

The IMA puff strategy created an error chain that has run amok with empty calories and hidden hunger that result in micronutrient deficiencies. Incredibly, this error has not been corrected in 70 years.

Human health impact

The Green Revolution focus on protein and calories created unintended consequences that have been catastrophic for several billion people, many of them children.

Agri-policy compensates farmers for crop yield, (tonnage), and ignores the protein quality, nutrient density and the diversity of vital nutrients needed to support healthy children, pregnant mothers and other adults. Food processors compound farmers' errors by removing natural fibers and inserting carbohydrates, sugar, fats and salt.

Farmers produce for yield, based on weight, with no nutritional consideration. GMO seeds are genetically modified to produce even more weight, which is predominately water. Increased crop weight often dilutes nutritional density. The USDA reports over 90% of the corn, soy and cotton grown in the US are monocultures that use GMO seeds.[87]

Western Diet of Processed Fossil Foods Manufactured Foods with Hidden Hunger		
Processed foods	**Food types**	**Chronically low**
• High fat	• Highly processed	• Low fiber
• High cholesterol	• High animal dairy	• Low nutrient quality
• High protein	• High animal meat	• Low nutrient density
• High sugar	• High GMO	• Low bioavailability
• High salt	• Pesticide residuals	• No bioactive compounds

The Green Revolution

Modern farmers produce nutrient deficient, calorie-dense food, above, that shifts consumer costs from food to healthcare. Food costs in the US are about 9% of living expenses, but healthcare costs have doubled in the last decade to 18%.[88]

Food policy should work to ensure a plentiful supply of healthy food. But the current system of agricultural subsidies mostly benefits large-scale growers of commodity crops such as corn and soybeans. Our diet is dominated by processed foods made from these subsidized crops. A growing body of research connects this diet to increases in obesity and the many diseases that go with it.
– Union of Concerned Scientists[89]

Hidden hunger

Why are children and family members not getting good nutrition? The root causes are hidden hunger from nutrient deficiencies and low staple food prices. Fossil foods suffer from hidden hunger, which results in foods with empty calories.[90] The term "hidden" refers to produce that shows few visible nutrient deficiency symptoms, (Figure 3.3.)

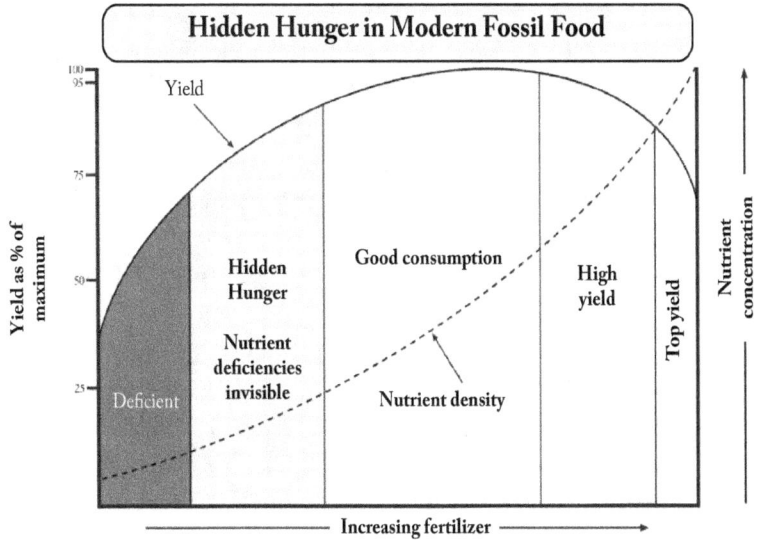

Figure 3.3 Hidden hunger diminishes nutrition in modern foods

29

IMA staple crops lack of one or more nutrients, particularly iron, magnesium and zinc, as well as essential vitamins, especially Vitamin A. Hidden hunger inflicts severe disorders on consumers.[91]

The nutrient content may be above the deficiency symptom zone needed for passable appearance, but below the zone for optimal crop health. IMA farmers grow the same crop year-after-year, to maximize profits. Predictably, systemic extraction strips out vital soil nutrients.

The appearance of produce with hidden hunger can be deceiving. Field tomatoes look puffed up today compared with 30 years ago, but the taste, texture and nutritional value has degraded by 40%.[92]

Vegetables, grains and fruits need a full complement of micronutrients to incorporate them into the fruit of the vine. When micronutrients, minerals, vitamins and trace elements have been extracted from the field and not replaced, the crop suffers from hidden hunger. Clear proof comes from the taste difference between heirloom and field tomatoes.

Crops allow farmers a wide zone, where withheld nutrients (fertilizers), show no visual effects. Farmers know that extra nitrogen creates larger produce, even when multiple micronutrients are absent. Consumers are attracted to larger produce. Few consumers suspect that the extra weight comes primarily from water; not nutrient-rich biomass.

Farmers are paid by weight, not nutrition. Consequently, many fossil foods are nutrient deficient because farmers hold back fertilizer, especially micronutrients, to save money. Most farmers do not replace micronutrients to fields because those formulations may not be available.

Applying only macro fertilizers, NPK, may represent 30 to 40% of the cost of the crop, which makes avoiding the addition of micronutrients to fields sensible from a cost perspective. Hidden hunger in food transfers directly to hidden hunger in people, called nutritional deficiencies. The lack of micronutrients in the fossil foods results in empty calories that deliver few nutrients per bite.

Hidden hunger imposes a painful toll at every stage of life.

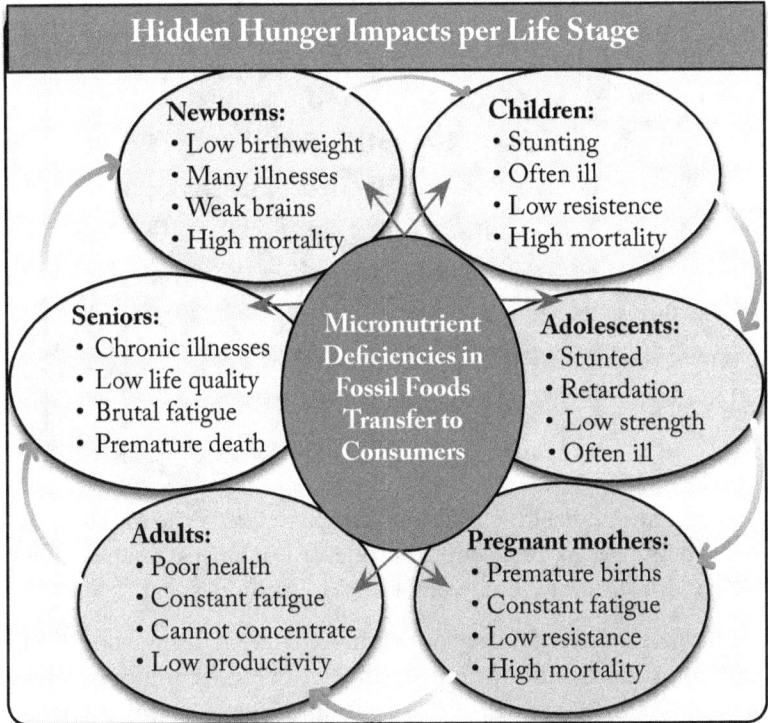

Impacts from hidden hunger caused by micronutrient deficiencies

Food processors amplify the "puff them up" strategy by adding more calories for energy. They remove natural fibers and add extra sugar, fat and salt, which makes food attractive but expands the consumption of calories devoid of phytonutrients.

While produce with hidden hunger may not show visible effects, hidden hunger in people results in child stunting and wasting, anemia, obesity and diabetes and heart disease. The Western diet of fossil foods lacking in micronutrients impairs natural immunity and increases disease in every major body system, metabolic diseases, diabetes and cancers.

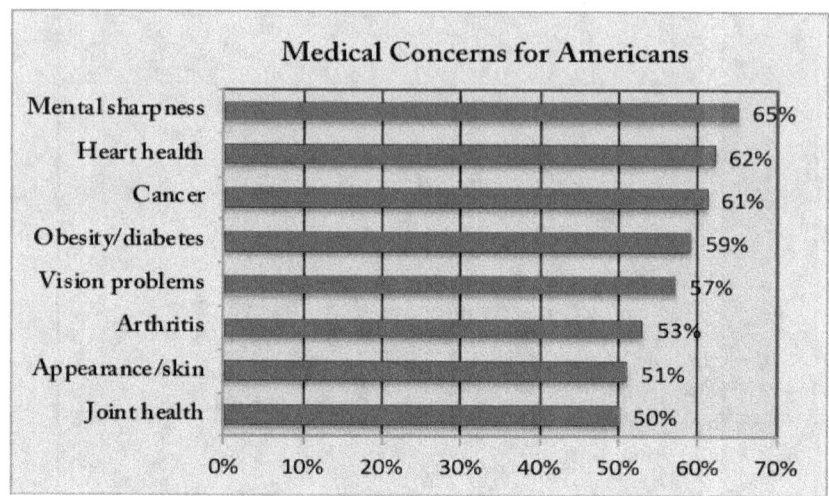

Medical Concerns for Americans

Concern	Percentage
Mental sharpness	65%
Heart health	62%
Cancer	61%
Obesity/diabetes	59%
Vision problems	57%
Arthritis	53%
Appearance/skin	51%
Joint health	50%

The cost of chronic health conditions associated with the Western diet and hidden hunger is not only imprudent, but unsustainable. These chronic conditions cost US citizens over $3.3 trillion a year.[93]

Over half of adults over 50 have one or more chronic health condition.[94] About 25% of adults suffer from two or more chronic conditions. Seven of the top 10 causes of death are chronic diseases. Micronutrient deficiencies are directly linked to chronic diseases.

Low staple food prices

How can low food staple prices block access to healthy foods? Food staples are priced artificially low due to USDA deceptive subsidy math. Fossil foods are propped up with billions in Big Ag subsidies, which reduce their true market price by more than 50%.

Consumers on tight budgets select the cheapest foods, which transfer hidden hunger and deliver empty calories. Cheap fossil foods that populate the Western diet make whole foods expensive in a relative sense. Fossil foods should require labels that informs consumers that they will face grave health hazards by eating these foods. Freedom food labels, coming soon, will expose the detrimental impacts of fossil foods.

The Green Revolution

Summary

IMA may have reached peak production. The planet simply has insufficient economically recoverable resource reserves to increase food production by the 70% the FAO projects to be needed in 30 years.

The Green Revolution has not resolved world hunger as today more people than ever are malnourished and hungry. Modern food production and processing have combined to create a puff strategy of cheap calories deprived of micronutrients. The predictable result: an epidemic of obesity, diabetes heart disease and cancers. The Western diet of fossil foods costs trillions in medical care but is even more costly in its appalling social impacts on families, education and healthcare.

IMA practices are unsustainable because many of the assumptions made about the Green Revolution turned out to be in error. The unintended consequences of those errors and the violence these false assumptions have inflicted on people, producers and our planet are examined in the next chapter.

4. False Assumptions Underlying Fossil Foods

The IMA designers made catastrophic errors that have had dire consequences on people, producers and our planet.

Key takeaways

- What are the false assumptions underlying the Green Revolution?
- What are the consequences of those errors?

The Green Revolution developed IMA production for fossil foods based on a series of assumptions. The bad news – most the assumptions turned out to be false. The good news – the Emerald Renaissance architecture addresses each of the errors with credible solutions.

Green Revolution -   Fatal Errors for Fossil Foods
1. Protein resolves huger; nutrition is not important.
2. Micronutrient deficiencies are unimportant.
3. Agri-productivity increases are sustainable.
4. Natural resources are plentiful and exploitable.
5. Fertile cropland is cheap and practically limitless.
6. Mechanical methods are superior to nature.
7. Fossil fuels make a solid foundation for food.
8. Roundup, glyphosate cocktails do no harm
9. Food miles are not relevant.
10. Soil microbial life is expendable.
11. Ocean fisheries offer endless catches.
12. Deforestation is necessary to produce food.
13. Staple food production trumps biodiversity.
14. Monocultures are smart; lower production cost.
15. Systemic nutrient extraction has no consequence.
16. Pesticides/poisons make sense to protect crops.
17. Agri-pollution of air and water is manageable.

False Assumptions Underlying Fossil Foods

Possibly the most critical mistake was defining global hunger as a protein crisis, which led to the focus on protein, calories and yield weight at the expense of nutrition. The ill-fated consequences have led to expanded hunger and malnutrition, plus our obesity and diabetes epidemic.

Additional Green Revolution assumption errors also have robbed our children of their natural resource stores and destroyed the croplands and waterways our children will need to grow fossil food.

Other critical false assumptions

1. Plentiful natural resources are a gift from nature and ripe for extraction and our unfettered consumption.

Greed may have driven this assumption. Natural resources probably seemed limitless in 1950. Fossil resource shortages, hidden hunger, environmental pollution and global warming were not table topics.

The IMA designers built the model on an eroding foundation of fossil resource consumption – fertile soil, fresh water, fossil fuels, fertilizer and pesticides and poisons. Systemic extraction and consumption made sense when resources were plentiful.

Unfortunately, not only are natural resources becoming increasingly scarce and expensive, but IMA's enormous waste and pollution are destroying our climate, fouling our atmosphere, poisoning critical waterways and decimating croplands and ecosystems.

Food production consumes about 50% of the planet's land surface globally, and about the same amount in the US.[95] Waste and pollution have degraded and destroyed much of the cropland needed for food production. IMA, from field to fork, consumes 32% of the world's total fossil energy, which is unreasonable and unsustainable.[96]

Enormous water consumption has extinguished aquifers and reservoirs, forcing many farmers to abandon their farms and move their families. Some Agri-chemicals such as phosphorus are becoming increasing scarce and expensive. Phosphorus price rose nearly 700% over a recent 18-month period. Agri-chemicals affordability and availability threaten

False Assumptions Underlying Fossil Foods

the future of global food production. Arizona State University's excellent Sustainable Phosphorus Alliance explains how scarce P has become.[97]

The Vital Four natural resource model, PAWS, put society at high risk for disastrous food production loss.

Phosphorus		P lies at the center of food universe. No cellular metabolism happens without P. Only a few countries have P mines and several mines have gone extinct.
Air		Air pollution costs over 2 million US life-years lost a year. Farm emissions block sunshine, which reduces crop yields. Farmers suffer from multiple respiratory diseases.
Water		While aquifers crash, cities demand more water. Farmers fight a losers' game for water rights. US farmers irrigate 55 million acres.
Soil		Huge topsoil loss and degradation leaves farmers with systematically lower yields and higher hidden hunger in crops.

Regional scarcity, high prices or outages of any of the viral four resources spells disaster for IMA farmers and American consumers.

2. Fertile cropland is cheap and practically limitless.

Unfortunately, IMA practices cause the fertile topsoil needed for food production to escape from fields with wind, rain and irrigation.

Globally, farmers have had to abandon more cropland, 2 billion hectares, than are now under cultivation for crop production, 1.5 billion hectares.[98] Farmers do not abandon their cropland easily. When fields become unproductive or sufficient water fails, they have to leave their farm. The farm family typically must migrate too.

Fertile cropland is precious, but increasingly less fertile. About 80% of the world's agricultural land suffers moderate to severe erosion.[99] Worldwide, erosion on cropland averages about 30 ton/ha/year but a single storm can remove 400 ton/ha overnight.[100] Erosion carries away fertile topsoil at over 20 times the natural replacement rate.[101]

Globally, about 75 billion tons of fertile soil are lost from food production systems each year.[102] Soil erosion and nutrient depletion from repeated crops destroys fertility. Land degradation, largely from agriculture drives species extinctions, intensifies climate change and contributes to mass human migration, conflict and wars.

3. Mechanical and synthetic technologies provide a superior alternative to the limits of nature.

IMA uses loud and violent machines to plough, cultivate and harvest.

Mechanical technologies were critical in winning WWII. Big machines proved their worth in war and seemed far stronger than the limits of nature. Ever-larger machines reduced human labor substantially.

Ironically, the idea that technology provides a source of plenty led to a series of mechanical tools that created new scarcities though massive resource consumption, excessive waste and ecological destruction.

4. Fossil fuels are plentiful and cheap and provide a solid foundation to produce food.

This probably seemed logical at the time when diesel was 10 cents a gallon. IMA foods require over 10 times the energy input as they return in calories for consumers. IMA production from farm to fork contributes 40% of all GHG to the atmosphere, which amplifies climate violence. Fossil fuel dependency drives up food costs in lock-step with the increasingly costly fossil fuels and reduces food access for the poor.

5. Food miles are not relevant.

Most IMA production occurs great distances from cities. Modern processors apply stabilizers, preservatives and other food processing tools to extend shelf-life. Food miles were not relevant when diesel fuel was cheap, cities were not contaminated with black soot particulates, and our atmosphere did not suffer from the heavy burden of warming GHG.

6. Soil microbial life is expendable.

IMA aggressively destroys beneficial microbial communities. Giant equipment crushes soil, then while cultivation slashes turns over the soil, killing microbial life. Fertilizers and pesticides inflict poisons that exterminate the microorganism communities that would otherwise work in symbiosis to enhance soil fertility.

Soil health depends on the precious organic component that house many living creatures that support crops. An acre of living soil may contain 900 pounds of earthworms, 2400 pounds of fungi, 1500 pounds of bacteria, 133 pounds of protozoa, 900 pounds of arthropods and algae, and possibly some small mammals.[103]

An acre of soil may contain over 10,000 species of microorganisms, which contributes to biodiversity and supports plant growth.[104] IMA practices decimate microbial communities and destroy diversity.

Soil organic matter may be the smallest but most critical component for crops. Organic matter powers soil biological, chemical and physical properties and consists of raw plant residues and microorganisms, (3-10%); active organic traction, (10-40%); and resistant or stable organic matter, (40-60%) called humus.[105]

Modern farmers replace the NPK macrofertilizer extracted by each crop but they typically replace neither the micronutrients nor the humus removed by each crop. The predictable result – fertility and crop productivity systemically diminish, along with nutrient dilution and hidden hunger.

7. The oceans offer plentiful fisheries and can supply endless catches.

Industrial fishing mirrors IMA with heavy emphasis on mechanical solutions while turning a blind eye to natural biological systems.

Bottom trawling is the ocean equivalent of clear-cutting forests. Ships drag enormous, heavy nets over the seafloor to catch fish. Trawling destroys all life, including deep sea coral and sponges, and other sensitive seafloor life. Nets are indiscriminate. For every ton of prawns caught, three tons of little fish are caught in the prawn nets and thrown away.

In only 55 years, industrial fishermen have managed to wipe out 92% of the ocean's top predators, disrupting marine ecosystems.[106] These beautiful animals include sharks, swordfish, marlin, and king mackerel. Peak catch occurred in 1989, when 90 million metric tons of catch were taken from the ocean.[107] Global fishery yields have declined ever since.[108]

A study of catch data published in the journal *Science* predicts that if fishing rates continue apace, *all* the world's fisheries, every species of wild-caught seafood – from tuna to sardines – will have collapsed by 2048.[109] Collapse occurs with a 90% depleti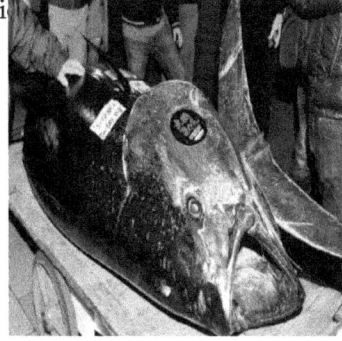

Big fish have become so scare that a Japanese restaurant owner, Kiyoshi Kimura, paid a record $3.1 million (333.6 million yen) for a 612 pounds (278 kg) bluefin tuna at the 2018 New Year's auction at Tokyo's Toyosu fish market. With most the big fish gone, industrial fishermen are now moving down the food chain.

8. Deforestation is necessary to produce food.

While industrial fishermen are removing food from the oceans, farmers are denuding forests to make more cropland. About half of the world's tropical forests have been cleared and all their stored carbon released to the atmosphere. Over 18 million acres (7.3 million hectares) of forest are lost each year.[110] Deforestation adds 15% to all GHG emissions.[111] These GHGs add to agriculture's 40% contribution to total emissions.

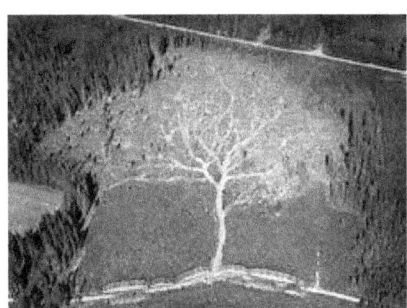

Forests contribute around 30% to atmospheric oxygen. The other 70% comes from algae, largely phytoplankton in the oceans. Deforestation disrupts the water cycle resulting in changes in precipitation, erosion and waterways. Trees extract groundwater through their roots and release it into the atmosphere. When part of a forest is removed, the trees no longer transpire this water, resulting in a drier climate. Deforestation

41

reduces the content of water in the soil and groundwater as well as atmospheric moisture.

A minimum of 60% forest cover is necessary to prevent serious soil erosion and landslides.[112] Removing plant cover eliminates the roots to hold the soil in place during heavy tropical rains. Rains wash away the topsoil that held the nutrients necessary to regenerate future vegetation. Replacement crops cannot hold the soil. Common crops such as coffee, cotton, palm oil, soybean and wheat typically accelerate erosion.

In some hilly countries such as Haiti, over 75% of cropland that replaced forests has been abandoned due to erosion. Slash and burn farmers that clear-cut a hillside may find the land holds fertility for only one or two seasons. Erosion forces farmers to move on and clear more forest, which continues the cycle of soil loss.

Land degradation, biodiversity loss and climate change are three different faces of the same central challenge: the increasingly dangerous impact of our choices on the health of our natural environment. We cannot afford to tackle any one of these three threats in isolation – they each deserve the highest policy priority and must be addressed together.

– Sir Robert Watson, Chair of IPBES[113]

Tropical rainforests are the most diverse ecosystems on earth and harbor about 80% of the world's known biodiversity Rainforest deforestation causes the loss of 137 plant, animal and insect species every single day, about 50,000 species a year.[114]

9. Staple crop biodiversity does not matter.

Today, 75% of the world's food is generated from only 12 plants and five animal species.[115] More than 90% of crop varieties have disappeared from farmers' fields. Half the breeds of many domestic animals are extinct. IMA animal production winnows biodiversity quickly as 30% of livestock breeds risk extinction and six breeds are lost to each month. Species extinction has accelerated to 1,000 times faster than expected, due to human actions such pesticide applications and deforestation.[116]

False Assumptions Underlying Fossil Foods

Extinct meat animals – Bos Premigenius, Manx Loaghtan lamb, Auroch

Climate change and IMA mechanical and chemical methods are predicted to drive 25% to 37% of all species to extinction over the next 30 years.[117] Since 1950, roughly 75% of plant genetic diversity has been lost.[118] IMA farmers have abandoned their diverse local varieties with unique tastes for genetically uniform, higher-yielding varieties.

Lack of crop biodiversity places the global food supply at severe risk. One pest vector could wipe out an entire crop. Plant genetic diversity is critical not just for taste, nutrition and variety, but also for beneficial insects. When certain crops are lost, pollinators, other insects and microbes are also likely to become extinct.

10. Monocultures make sense.

Monocultures have become the foundation for IMA. Farmers grow a single staple crop intensively on a very large scale, such as corn, wheat, soybeans, cotton and rice. Monoculture farming relies on enormous chemical inputs, including synthetic fertilizers and pesticides.

Monocultures are aptly called *"The curse of a commodity business."* Focus on yield led to all the economics and subsidies tied to tonnage, not nutrition. Production-focused agriculture fails to deliver healthy food.

More than 33% of US farms are larger than 2,000 acres. Mega farms are richer, too as half of farm production comes from farms with annual sales of at least $1 million. Most of the government generosity in subsidies, crop insurance and commodity support prices are reaped by mega farms.

A recent USDA report concluded acknowledged megafarm problems.

False Assumptions Underlying Fossil Foods

Large-scale farming operations force small farms out of business, damage the viability of rural communities, reduce the diversity of agricultural production, and create environmental risks through their production practices.[119]

Monocultures have profound climate and environmental implications. Mega-farms systematically degrade cropland from the combined effects of erosion, cultivation, crop nutrient extraction and agri-chemicals. These actions directly increase emissions of greenhouse gases.

Monocultures drive biodiversity extinction and create risk for crop failure

Fertilizers are needed because growing the same crop in the same field year after year quickly depletes soil nutrients. Pesticides are needed because monoculture fields attract and harbor weeds and insect pests.

Commodity farming the duoculture, corn and soy, discourages environmentally sensible practices like no-till farming and crop rotation that take CO_2 from the air, store it in the soil and improve soil health.

IMA waste erodes and poisons flora, fauna, beneficial insects and aquatic creatures. Monocultures put our entire food supply at risk for a single disaster, disease, pest vector or economic turbulence.

Repeatedly growing the same staple crop creates a recipe for disaster from invasive weeds and insects that gain advantage season after season. Monoculture farmers clear out natural ecosystems that harbors biodiversity for expanded planting. IMA practice eradicates biodiversity in crops and in ecosystems.

Monocultures decrease the variety of crops grown, which not only reduces the array of consumer food choices but also eliminates many specialty pollinators, beneficial insects, butterflies, bats and birds.

A 2019 meta-analysis of the 73 best insect studies concluded that insects are hurtling down the fast lane to extinction. Over 40% of insect species are threatened with extinction. Unless we change our ways of producing food, insects will be extinct in a few decades. Insect population loss to our planet's ecosystems will be catastrophic. Insects are pollinators and nutrient recyclers. Ecosystem services provided by *wild insects* contribute $57 billion annually in the US.[120] IMA and agri-chemical pollutants are the main driver of insect destruction.

Monarch butterflies experienced an 86% die-out in 2018

Industrial-scale, intensive agriculture destroys insect habitat and entire ecosystems. IMA eliminates all trees and shrubs that normally surround fields. After cultivation, only bare land remains and fields are repeatedly treated with synthetic fertilizers and pesticides, both of which kill insects and soil microbes. New, powerful insecticides, including neonicotinoids and fipronil, sterilize the soil killing all life indiscriminately. Agri-poisons persist in the soil and migrate into nearby ecosystems.

When the beautiful flycatcher finds no flies, she dies. Insect loss means food loss for birds, reptiles, amphibian and fish populations. Without insects, all these animals starve to death. The authors conclude *the world must change food production methods.*[121]

False Assumptions Underlying Fossil Foods

Monoculture staples are easier to grow and produce more total yield, but they limit farmers to the commodity price for the single product. The USDA reports that 88% of the corn and 94% of soybeans are GMO monocultures.[122] In 2018, soybean farmers made all their agri-input investments, planted in April and watched the futures price of their single commodity fluctuate with normal market forces. Soybean futures crashed due to the US/China trade war. Many farmers decided it was not worth their time and cost to harvest their fields.[123]

Monoculture crop failure and bankruptcy constantly threaten farmers

Monoculture production costs rise with fossil fuels. The cost of producing an acre of soybeans has risen 273% since 1975, and currently costs the average US farmer $443 per acre.[124] The cost of GMO seeds has risen 600%. Net income for US soybean farmers was just $31 per acre in 2017 and is likely to be even lower in 2018 due to the self-inflicted political tariff war.

Mega-mergers of agricultural chemical and seed companies—Monsanto and Bayer, ChinaChem and Syngenta, Dow Chemical and DuPont—have concentrated seed and agri-chemical technology in the hands of a few companies. Farmers have lost options over what to plant and how.

False Assumptions Underlying Fossil Foods

11. Pesticides and herbicides improve farming

IMA farmers pre-treat their fields with the weed killer, Roundup, to kill emerging weeds that would compete with their crop. Weeds are hardy. Weeds are quickly becoming resistant to Roundup, so farmers often spray a field several times with glyphosate.

Public health scientists and the EPA in 2017, found most food and drinks, including drinking water are contaminated with glyphosate. The WHO determined in 2015 that glyphosate is a probable carcinogen.[125] California concurred in 2017. The US PIRG Education Fund tested 20 wine and beer samples – 19 contained glyphosate.[126] Cow, goat and human breast milk tests were similar.

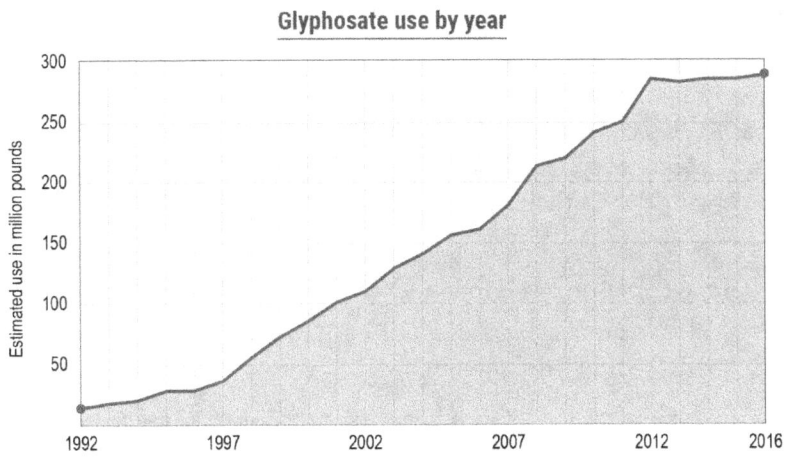

Glyphosate use by year

Source: U.S. Geological Survey

Roundup is a poison cocktail that includes glyphosate, in endocrine disrupter. Roundup has a number of devastating biological effects.

- Nutritional deficiencies – glyphosate immobilizes certain nutrients and alters the nutritional composition of the treated crop.
- Increased toxin exposure, glyphosate and formaldehyde in foods.
- Neurodevelopmental, reproductive, and transgenerational effects.
- Disruption of the biosynthesis of aromatic, essential amino acids,

47

- Gut dysbiosis, imbalances in gut bacteria, inflammation, leaky gut, and food allergies such as gluten intolerance.
- Sulfate deficiency, poor sulfate transport and sulfur metabolism.
- Systemic toxicity – extreme disruption of beneficial microbial function throughout the body, allowing overgrowth of pathogens.
- Enhancement of damaging effects of other food-borne chemical residues and environmental toxins as a result of glyphosate shutting down the function of detoxifying enzymes.
- Creation of ammonia, which can lead to brain inflammation associated with autism and Alzheimer's disease.[127]

Despite high risks, Roundup use expands. Glyphosate sprayed every year could apply 0.8 pounds on every cultivated acre of land in the US, and half a pound on every cultivated acre of land in the world.[128]

Several sources have found herbicide residues in organic produce, but at lower levels than IMA food and drinks.[129] Glyphosate contamination is nearly ubiquitous, including ice cream.[130] Agri-poison residues are problematic due to their severe health risks. In addition:

- The EPA does no regular testing and has not set acceptable levels for agri-poison levels in US food or drinks.
- Water treatment plants do not have the equipment to clean agri-poisons out of drinking water.

The first court trial over whether Monsanto's Roundup causes cancer ended in 2018 when a San Francisco judge upheld a jury's verdict that the weed killer did make DeWayne Johnson sick with non-Hodgkin lymphoma. The judge cut the jury award from $289 million to $78 million. Thousands of others, mostly farmers, are now alleging that their incurable cancers may have been caused by Roundup. A second jury also concluded that Roundup causes cancer. Bayer/Monsanto are likely to pull Roundup from the market as jury awards mount.

After pre-treatment with herbicides, farmers apply more poisons, pesticides and fungicides. The economic toll of pesticides residuals in fields, on food and in packaged food is astonishing and deadly. Pesticides create an urgent public threat posed by endocrine disrupting chemicals, especially to our children. Pesticide residuals disrupt the body's hormones and are linked to a myriad of severe health problems, such as impaired brain development, lower IQs, behavior problems, infertility, birth defects, obesity and diabetes and all the follow-on diseases.

Toxic heavy metals attach to major organs and interrupt normal organ functions. Lead poisoning in Flint Michigan focused world attention on the effects of heavy metals poisoning. The Flint lead problem affected 100,000 people.[131] Persistent pesticide residue exposure affects at least 10 times more people nationally and 100 times more people globally.[132]

A recent study led by Pete Myers, founder of Environmental Health Sciences, calculates that pesticide costs the US more than $45 billion annually from health care costs and lost wages.[133] Pesticide exposure causes an estimated 2 million lost IQ points and another 7,500 intellectual disability cases annually.[134] The calculations are made from metrics developed by the Endocrine Society, WHO and the UN Environment Program.[135] Additional illnesses will add to the pesticide drag due to the lag effect in heavy metals exposure. Many illnesses such as cancers, neurological, brain and respiratory diseases from toxic chemical exposure appear years after exposure.

Pesticides severely degrade the environment too. The Geological Survey (USGS) and Fish and Wildlife Service (FWS) found that pesticide endocrine disruption causes sex changes among small and largemouth bass. Males had female eggs inside their testicles.[136] Incredibly, IMA continues to increase applications of agri-poisons that put all of society at terrible health risk, especially to unborn children.

12. The Farm Bill helps Americans

No, the Farm Bill benefits Big Ag. The Farm Bill relegates small and medium-sized farmers, consumers and nutritionists to background noise. The Farm Bill amplifies the wrong priorities and degrades health to people, farmers, croplands, wetlands and waterways.

The Farm Bill benefits immensely from the early Iowa Caucus. Presidential candidates know their political platform must align with Big Ag or their Presidential bid will be toast.

John McCain discovered this when he scuttled his 2008 Presidential hopes in the Iowa Caucus debates when he said he was against corn ethanol. He said correctly that it takes more energy to make than it delivers. In the same debate, McCain also said that corn-based ethanol production in the US would be totally financially impossible if it were not for massive government subsidies.[137] McCain was punished for his honesty. Americans have paid an enormous price ever since.

The early Iowa Caucus is the only reason the corn ethanol biofuel mandates exist. No other country in the world is stupid enough to use food, especially corn, for biofuels. (Sugarcane does make sense in Brazil.)

13.   **Farm subsidies are good and help farmers.**

Subsidies benefit primarily millionaire farmers by giving them more money. Corn, soybean and cotton subsidies cause farmers to grow those crops, due to the price supports, even though they know that repeated monocropping will destroy their cropland. Subsidies remove crop choice from farmers, many of whom feel like tenant farmers. Subsidies drive monocropping, which destroys biodiversity.

Subsidies are insidious. They hide the true cost of food from the US public. They are destructive to health and amplify hidden hunger.

Subsidies do terrible international damage by undercutting food prices for Canadian and Mexican and Central American farmers. US subsidies have forced thousands of farmers into bankruptcy because they could not sell their crops for a fair price. Much of the conflict over Trump's border wall is inflicted by subsidized US foods that undercut Mexican farmers.

False Assumptions Underlying Fossil Foods

Canada and Mexico, top America trading partners, both have cases before the World Court calling US food subsidies unfair practice.

Scenarios for $10 billion wipeouts

IAM methods, especially monocultures, put all society in jeopardy for predictable catastrophic wipeouts.

> *Monoculture crops are genetic siblings of each other. They grow together and – with a single threat vector – all die together.*

Each of four fossil food scenarios could cost $10 billion or far more. Adoption of abundance food cultivation methods will avoid these very expensive disasters.

Fossil food wipeout. Practically all fossil foods and drinks are contaminated with agri-poison residuals. Farmers use pesticides, herbicides, fungicides and other poisonous agrichemicals to protect crops. Unfortunately, a majority of those poisons leave the farm with wind and water and damage and destroy flora and fauna. Harvest removes crops from the field but not the agri-poison residue. When the first study concludes that agri-chemical residuals causes cancer, respiratory or neurological diseases, consumers will fear fossil foods. **Fossil food flight** will be fast and expensive for fossil producers.

Farmers may quit fossil farming for the obvious health threat to themselves, their family and community. DeWayne Johnson's $289 million cancer award for his exposure to Roundup will be followed by thousands of other plaintiffs. Each new claimant has considerable incentive to expose the substantial research that shows pesticide endocrine disrupters do serious damage to human health, especially during womb-life. Thousands of US babies are born with developmental disabilities from their mother's agri-poison exposure.

GMO food wipeout. GMO food staples make up a majority of US foods while many other are made with GMO ingredients. No current credible medical research has found direct harm from GMO edible material. Scientists expect a lagged effect in that GMO food impacts may take a decade or two to show harmful effect. When medical science concludes that GMO foods are unsafe, consumers will look for alternatives. Smart people today are looking for alternatives due to implied risk. The EU has banned GMO foods and declared GMOs unsafe.

Pest wipeout. A single invasive, pesticide resistant insect, parasite, mold, mildew, nematode or other threat vector attacks GMO corn or soy. The corn borer may lay eggs on the corn cob in the Spring. The eggs hatch and feed on the kernels all summer, destroying the crop.

Roughly 12% of the total US soybean crop is lost every year to pests or weather. Soybean cyst nematode, (right) causes the largest losses – 95 million bushels.[138] Annual losses to corn crops hover around 12%, plus or minus 3%.

A 12% crop loss may seem small, but it is heartbreaking to individual farmers. Losses typically occur in geographical clusters that share the same microclimate and pest vectors. Regional crop losses can devastate rural agricultural communities. The spring 2019 "bomb cyclone" that hit Nebraska and the entire Mid-West killed millions of livestock, flooded croplands and ruined stored grains. Many farmers will miss planting season because soils are saturated and too wet for equipment. Many farmers will go bankrupt from the single storm. Crop insurance covers some of the loss, but farmers many never recover their financial stability.

A recent study predicts the global banana industry could be wiped out in just 5 to 10 years by fast-advancing fungal diseases. The disease called Sigatoka, caused by three types of fungi, already destroys 40% of banana yields annually.[139] Similarly, the elm borer and the bark beetle are currently devastating millions of acres of non-food flora.

Temperature spike wipeout. A single early spring or late fall temperature spike or fierce storm could wipe out millions of acres of crop staples. Monocrops are particularly vulnerable for two reasons. These crops lack biodiversity, which means in essence they all die together. GMO crops are bred for yield weight, not weather resistance.

False Assumptions Underlying Fossil Foods

Temperature spikes and storms are more likely to destroy monocrops than diverse plantings.

Droughts and heatwaves in 2018 caused over $10 billion in agricultural losses across Europe and Australia. Between 2005 and 2015 natural disasters cost the agricultural sectors $96 billion in damaged or lost crop and livestock production.[140] Trendlines for annual IAM agricultural loss are increasing, which has forced many farmers into bankruptcy.

Fossil fuel wipeout. A single Gulf of Mexico storm could tangle diesel fuel production, refining and distribution for months, which would leave all IAM farmers in an untenable situation. Farmers would have invested all the agri-inputs into their crop, only to find there is no diesel for harvest, processing and transportation. Spot fuel shortages have left crops in the field in many areas of the world, most recently in Bolivia.

Infrastructure wipeout. Cities such as Las Vegas are extremely vulnerable to a single event – earthquake, wildfire, terrorism attack – that would block the single main road into the city. Cities typically have food stores for only 3 to 5 days. A breakdown of infrastructure would cause significant damage to people and the economy.

Natural resource cascade. IAM dependence on fossil resources make a fossil resource cascade likely. The run begins with stealth. A silent speculator buys all the available phosphorus (or other resource) in a certain geography, such as the corn belt. As the speculator quietly corners the market, whispers shimmy across the agri-scape. Farmers buy as much P as they can and begin hording. They know that without the fertilizer, they cannot grow crops. These actions create a feeding frenzy where the price of P goes far above the price farmers can pay and still break-even on their crop.

A natural resource cascade results in considerable P horded and not used. Additional speculators are likely to join the P-bubble and buy and hold, waiting for higher prices. Many farmers cannot plant because they do not have a reliable, affordable source of P. P makes a good choice for a resource run because plants cannot be fooled with a synthetic.

False Assumptions Underlying Fossil Foods

A natural resource cascade in a region could force hundreds of farmers into bankruptcy, resulting in billion-dollar losses. A more sinister scenario has been actively contemplated. Several very wealthy international speculators have made attempts to buy fertilizer mines in order to create a monopoly and force prices higher. Recent attempts were thwarted by countries declaring fertilizer a strategic natural resource. Fertilizer prices will continue to move higher as demand increases and supplies wane.

How could this error chain happen?

In the future, when the USDA Secretary must report to Congress about how USDA policies could have allowed a multi-billion-dollar tragedy to happen, the answer Secretary's predictable answer will be:

> *"We knew this loss was possible. We just didn't think it would happen."*

Avoid $10 billion wipeouts

Abundance methods offer a smarter path that can avoid these expensive wipeouts. Freedom foods use no agri-poisons and no GMO foods. Microfarms do not grow monocrops and are nearly climate independent, which eliminates many fossil food health, biodiversity and production risks. Abundance methods can recover and reuse phosphorus from waste streams, which avoids market price volatility.

The error chain of false assumptions created unintended consequences – increases in the food production costs, increasingly scarce and expensive natural resources and severe pollution to air, water and ecosystems.

> *What will our children do for food when the natural resources they desperately need for food production have become extinct?*

The next chapter explores IMA's ugly impacts on the health and safety of farmers and our environment.

5. The Green Revolution's Violence to Producers

Building a food supply system dependent on industrially fixed nitrogen and mined inorganic chemicals, makes our ability to feed ourselves dependent upon nonrenewable fossil fuels, limited agricultural chemicals and the wisdom, benevolence and cooperation of a few corporate executives that head the world's chemical fertilizer plants, mines and petroleum companies. These executives share a very poor history.[141]

Key takeaways

- What impact does IMA have on health and safety?
- What are critical environmental issues?

Sustainable systems should show positive impacts in the areas of health and safety, environment, social and economics. IMA is not sustainable because modern food production creates substantial health risks for people, animals, plants and our environment.

Health and safety

Industrial farming creates severe health risk for farmers, farm families, farm animals and their rural communities. IMA ranks among the most hazardous industries due to the use of heavy and loud machinery, heavy labor and agri-chemicals.[142] Farmers are at high risk for fatal and nonfatal injuries, lung diseases from dust and pollutants, hearing loss from heavy equipment, skin diseases, chemical-related illnesses, and cancers associated with chemical use and prolonged sun exposure.[143] In an average year, 516 US farm workers die and about 250 IMA workers suffer injuries, 5% of which are permanent impairments.

The Green Revolution's Violence to Health

Water pollution from fertilizer and pesticide runoff severely contaminates waterways, lakes and aquifers. In some corn-producing states, over 80% of the waterways are unfit for human recreation. In Big Ag states, nitrogen pollutes air and water and creates a smog of dangerous nitric oxides, ammonia and ozone, which can impair our ability to breathe, limit visibility and alter plant growth.[144] When excess nitrogen returns to the earth from the atmosphere, it can harm the health of animals, forests, soils and waterways.

Nitrate water pollution results in blue-baby syndrome, often with heart and breathing problems. Eutrophication in waterways occurs from agri-overspill from too many fertilizer nutrients. Eutrophication in US freshwaters costs are calculated at $2.2 billion annually, not counting estuaries, the Gulf of Mexico and other dead zones.[145]

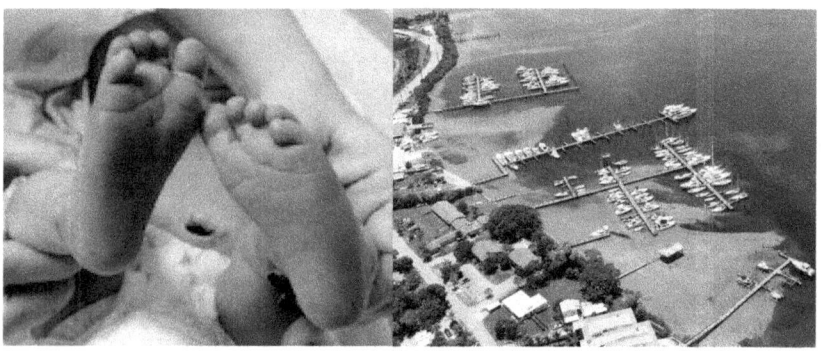

Blue baby syndrome and eutrophication – from nitrogen pollution

Animals raised for food produce approximately 130 times as much excrement as the entire human population.[146] Animal farms pollute our waterways more than all other sources combined.

Water pollution from fertilizer overspill has created over 400 dead zones globally that are expanding at 10% a decade. Freshwater pollution by phosphorus and nitrogen costs US government agencies, drinking water facilities and individual Americans more than $4.3 billion annually.[147]

Pesticides and poisons. Herbicides, fungicides and insecticides used in agriculture cause pregnant mothers to abort or bring to term infants with severe developmental disorders. These toxic compounds cause both acute and often life-threatening poisoning and long-term chronic illness for consumers, farmers, farm animals and rural communities. Pesticide residuals cling to produce, which amplifies health risks for unsuspecting consumers. The EWG provides a Dirty Dozen list of produce that carry the highest amounts of pesticide residue.[148] Ironically the top 12 list includes favorite foods recommended to improve health – apples, strawberries, peaches, spinach, grapes and tomatoes.

Antibiotic resistance. The USDA reported that animal agriculture consumes 80% of the antibiotics used in the US.[149] The overuse of antibiotics has accelerated the development of antibiotic-resistant bacteria, which cause 76 million cases of food-born illnesses a year and over 5,000 deaths.[150]

IMA displays an appalling eco-footprint, degrading and destroying the very ecosystems needed for food production.

Environmental degradation

How could anyone have the poor judgment to design a food supply system that degrades and destroys the very cropland and ecosystems on which it depends for production? Fossil food depends on cropland fertility, abundant fresh water, biodiversity and healthy rural communities. IMA actions systemically degrade these critical features.

Cropland consumption. Maize, (corn) occupies more US cropland than any other crop, spanning some 97 million acres – nearly the size of California.[151] Globally, maize consumes about 550 billion cubic meters of water annually, which is 8% of global water use for crop production.[152]

Corn production applies over 5.6 million tons of nitrogen each year through chemical fertilizers, along with nearly a million tons of nitrogen from manure. Tons of nitrogen fertilizer volatizes into the atmosphere. About 60% of the fertilizer and soil, washes into waterways, polluting ground and surface water and damaging ecosystems.

The resources devoted to growing corn and other IMA food grains are increasing dramatically. Between 2006 and 2011, the amount of cropland devoted to growing corn in America increased by more than 13 million acres, in response to energy policy decisions that increased demand for ethanol. These new corn acres came from over a million acres of farms growing other crops. Another 1.3 million acres came from converting grassland and prairie to corn, which destroyed waterways, wetlands and the ecosystem flora and fauna.[153]

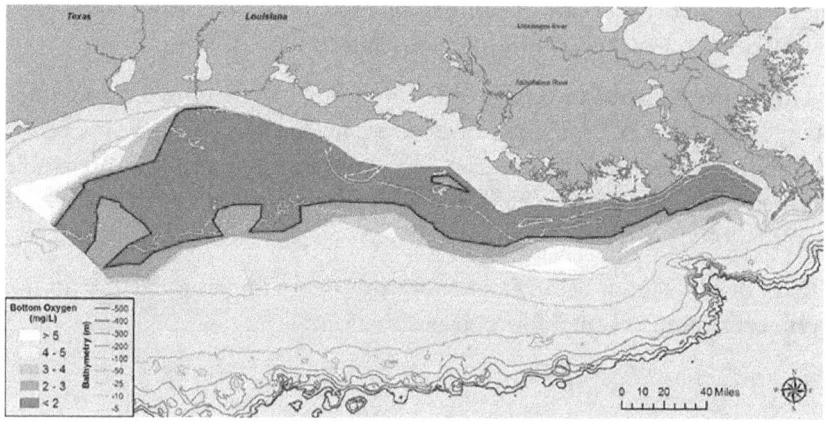

The Gulf of Mexico dead zone kills all aquatic life

Conversion of millions of acres to corn monoculture imposes substantially more natural resource consumption, pollution and loss of biodiversity in plants and animals.

Soil degradation. Monoculture farming the same land year after year depletes soil fertility by extracting soil nutrients and humus. Farmers replace only a few of the nutrients, which leads to hidden hunger in crops. Hidden hunger occurs with crops look good but lack critical nutrients because the nutrients were not in the soil.

WHO reports that farmers apply more than 200 million tons of chemical fertilizer globally. The use of chemical fertilizer destroys natural fertility by exterminating the beneficial soil microbes. Pesticides, fungicides and herbicides also poison beneficial soil microbes, further diminishing soil fertility. Over 5.6 billion pounds of pesticide are used globally, and 2.2 billion pounds in the US each year.[154]

IMA pesticides have killed 75% of flying insect populations in Germany over the last 27 years.[155] Pesticides threaten 4,000 species of bees, which pollinate over $20 billion in crops each year.[156] A recent count by the Xerces Society in California recorded fewer than 30,000 butterflies, an 86% decline since 2017.

Agri-poisons result in **poison resistant pests** that are expanding rapidly. IMA farmers need to continually use more fertilizers and pesticides each season to achieve the same results. Farmers in the US lost 7% of their crops to pests in the 1950s but their losses doubled in the 1990s even though more pesticides were being used.[157] Since 1945, over 1,000 species of pests have evolved a resistance to a pesticide.[158]

Cultivation and erosion. Monocultures degrade soil structure from above and below. Fields are left bare for much of the year. Farmers tend to cultivate cropland for planting at the same time spring winds blow topsoil from their fields. Heavy equipment cultivates and compresses the soil, leaving topsoil vulnerable to erosion from wind and water.

The International Food Policy Research Institute reports that each year an estimated 20 million hectares, (50 million acres) of cropland worldwide are abandoned due to soil erosion, irrigation salts and exhausted soil. The USDA estimates 5 million cropland acres must be abandoned in the US each year due the extinction of natural resources such as water or soil fertility.

Water consumption. Fresh water acts as a prime limiting factor for productivity for all terrestrial plants. Crops require enormous quantities of water for growth. Each hectare of corn transpires about seven million liters of water, (1,850,000 gallons) during the growing season and lose an additional two million liters of water by evaporation from the soil.[159]

The Green Revolution's Violence to Health

A recent comprehensive study, *The Water Footprint of Humanity*, found that IMA consumes an incredible 92% of available fresh water in the US.[160] Irrigation water must be clean, affordable and delivered on time. Over 46% of US surface waters are too polluted from agriculture for fishing, swimming, or edible aquatic life.[161] NOAA estimates that 80% the pollution to the marine environment comes from agriculture.[162]

Sprinkler and flood irrigation lose half the water to evaporation

Pumping from fossil aquifers increases energy costs as the wells must be drilled deeper. Deeper wells also increase the risk of heavy metals contamination such as lead and arsenic, that crops assimilate. Many of the critical aquifers used of food production globally will go dry in the next generation and farmers will be forced to leave their land.

IMA staple crops such as corn consume water on a massive scale. Corn requires three acre-feet of water from rain or irrigation during a growing season in Iowa but may require twice that amount in hot climates such as the US Southwest.[163] An acre foot, 326,000 gallons, covers an acre one foot high.[164]

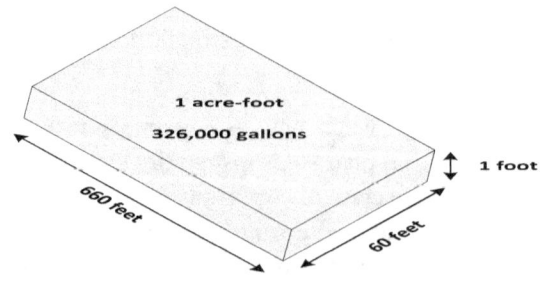

One acre-foot of water = 326,000 gallons

The Green Revolution's Violence to Health

The EPA estimates a family household uses 44% of an acre foot of water a year, which means an acre foot supports 2.3 households with a family of four. Three acre-feet, about 1 M gallons of water, supports one acre of irrigated corn or about seven families.

With conservative assumptions, a corn field consumes 2 million gallons per acre. A 1,000-acre corn farm may produce 1,400,000 bushels of corn while using the water equivalent to a city of 140,000 households.

The water calculation for ethanol ignores the water used by the farm families and water used in refining ethanol. Given the USDA average corn yield of 140 bushels per acre, an acre of corn produces:

140 bu/acre * 2.5 g ethanol per bu = 350 gallons of ethanol per acre[165]

and 1 M gallons of water per acre / 350-gal ethanol per acre

= 2,857 gallons of water per gallon of ethanol

Since 1g water = 8.35lbs, then 2,857 g * 8.35lbs = 24,006lbs = **12 tons**

3,000-gallon water tank

The 3,000 gallons or 12 tons of water required for each gallon of ethanol represents a water-thirsty way to produce low-grade energy. The water must not be too saline and cannot be recovered or reused. Irrigation water extracted from fossil aquifers typically cannot be replaced.

It seems inconceivable that corn would consume 12 tons of water to produce each gallon of ethanol. Another way to validate ethanol's incredible water cost uses the USDA ethanol yield per ton of corn, 89 gallons, and the approximation that producing one ton of grain requires 1,000 tons water.[166]

The Green Revolution's Violence to Health

This computation ignores the water cost of a rotation crop, families on the farm and the refinery water. This model yields 11.2 tons of fresh water consumption per gallon of ethanol. The US produced 14 billion gallons, (52.6 billion liters) of ethanol in 2011. The 168 billion gallons of freshwater consumed producing ethanol each year cannot be replaced and provide less than 1% of net US petroleum needs.

Waste biomass creates enormous labor, cost and waste problems. A farmer must cultivate, plant, tend and harvest a bulky biomass that yields only a tiny fraction of food. Corn requires 3 to 4 months to mature the roots, stalk, leaves, tassels, corn cob and kernels. The only edible biomass is the fruit of the grain, kernels, which yield a tiny amount of food.

A fresh cornstalk dug from a mature cornfield in Arizona demonstrates the meager food metrics.

Ann Ewen holds a corn stalk that measures 10' 1" tall and weighs 4 pounds. Field corn produces only one cob of 640 kernels. Ann holds the kernels that weighed 4 ounces, which represent 6% of the plant's weight and yielded only 4 ounces of protein.

A 6% biomass food energy yield may seem miniscule for such a large plant, but there's more bad news. Most of the kernel biomass, 63%, is non-food cellulose and water, which adds to processing cost and waste, but decreases food value. The net food biomass, 2.2% yielded a minute 1.2 ounces of protein. There is still more to this food story.

Corn, similar to most other terrestrial grains produces protein, but the protein is incomplete – lacking in several essential amino acids.

The Green Revolution's Violence to Health

Professional nutritionists recommend corn be paired with brown rice, couscous or cooked lentils to acquire the other essential amino acids for balanced protein.

Food grains require IMA farmers to make all the agri-input investments to grow a product that may produce a bit more than 2% food. Then the farmer must manage the massive waste biomass. Farmers must grow, fertilize, irrigate, harvest and handle a huge crop biomass where 97% of the plant production becomes waste. The waste biomass to food metrics for animal farmers, both meat and dairy, are even worse than food grains.

The use of corn for biofuels creates an even starker picture for biomass waste and pollution than corn for food. Over 40% of US corn and 30% of soybean production in 2018 was consumed to produce ethanol.[167] Each large corn plant yields less than 2.2 ounces of ethanol. Ethanol yields only 64% as much energy and gasoline, which drops the net yield to 1.5 ounces of gasoline equivalent energy. A Hummer using E-100, (pure ethanol) consumes about 16 corn plants per mile and burns 5,200 corn plants with each fill-up.[168]

IMA agri-policy, controlled by the relatively few agri-barons that won the Green Revolution, benefits those large agribusinesses but neither society nor the environment. Current agri-policy practically eliminates the IMA food productivity in grain crops such as corn and soybeans. Agri-policy sends these crops to make biofuels, not food.

An average Iowa cornfield has the potential to produce 15 million calories per acre each year. The single cornfield acre could sustain 14 people with 3,000 calories-per-day diet. However, the current agri-policy allocates 40% of corn to ethanol and 36% to animal feed. This leaves each corn acre producing only 3 million calories of food per year for people. This is only enough left-over calories to sustain three people per acre. This net food production is lower than the average delivery of food calories from poor farms in Bangladesh and Vietnam.[169]

Nitrogen pollution. Every single living cell – plant, animal or human – requires nitrogen, N, for its physical structure, function and reproduction. The atmosphere is made up of 78% N, but plants cannot use atmospheric N directly. N must be converted to nitrates or ammonia, which often occurs by N fixing cyanobacteria, blue-green algae.

Production of reactive N has increased 15 times since 1960, to 115 billion tons in 2017.[170] Over 80% of this N is used in IMA.[171] Synthetic N production manufactures over double the N of all natural processes on land combined. In the US, people consume only about 10% of the N farmers apply to their fields every year.[172] The remainder volatizes into the atmosphere or runs off into waterways.

Fertilized soils release more than two billion tons of greenhouse gases every year, especially CO_2, methane and NO_x, (N oxides). A recent scientific report of nine global environmental challenges that may make the Earth unfavorable for continued human development identified N pollution as one of only three problems – along with climate change and loss of biodiversity – that have already crossed a boundary that could result in disastrous consequences if not corrected.[173]

The excess reactive fertilizer is mobile and rides on air and water from the field. Additional N in the form of nitrate seeps into drinking water, where it lurks as a poison that can damage internal organs.[174] High levels of nitrates in drinking water have been linked to blue baby syndrome, when a baby's blood can't carry sufficient oxygen, as well as miscarriages.[175] A review by the National Institutes of Health shows elevated nitrate concentrations in drinking water raise the risk of cancer, Alzheimer's, diabetes and heart disease and drives up mortality rates.[176] Nationwide, utilities spent $4.8 billion to remove nitrates from public drinking water supplies in 2011.[177]

In the heart of the corn belt, Iowa's largest sewage treatment plant spends millions on nitrate removal. A 2007 Iowa Natural Resources report indicated 274 Iowa waterways were seriously polluted.[178] Studies in all the corn-belt states have produced similar results. A majority of waterways and well-water in the corn-belt are contaminated with fertilizer. Fertilizer run-off causes such a problem that Iowa was forced to install the largest and most expensive in the world. The Des Moines plant now must invest another $19 million in new equipment to reduce the fertilizer found in the state's waterways and well water.[179]

The Green Revolution's Violence to Health

Rural families not connected to the city's water treatment plant must depend on well water, which puts their families at substantial health risk, because their water contains carcinogens and other deadly poisons. A University of Iowa study tested 475 private wells in 93 counties. The unsettling results showed 49% tested positive for nitrates, 48% had arsenic and 43% tested positive for total coliform bacteria, which can make people extremely sick.[180]

Large amounts of nitric oxide volatize into the atmosphere as ammonia, where it creates smog and causes respiratory disease. Agriculture releases the largest human caused source of nitrous oxide, a highly reactive form of N that contributes to global warming and reduces stratospheric ozone that protects from ultraviolet radiation. Ecological N pollution in waterways also enables the spread of invasive species such as ragweed that elevates pollen pollution, mosquitoes and snails that carry disease.[181]

IMA's social and economic impacts are equally destructive.

Social and economic impacts

Social justice. After health, the most important goal for food production should be social justice, to assure everyone has access to good food. IMA fails this goal.

Food production per section of land has not kept pace with population growth for decades. Per-capita cropland production has fallen by more than half since 1960. Per-capita production of grains, 80% of what people eat, has been falling globally for 20 years.

Crop scientists have known for decades that plant diversity provides many benefits for soil fertility and for crops. Social scientists have proven that diversity in any industry provides a host of benefits. IMA cannot approach food justice unless women and ethnic minorities are empowered to grow crops.

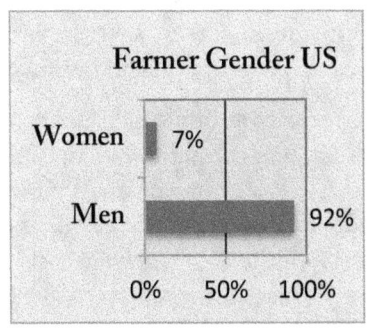

Farmer Gender US

Women 7%

Men 92%

0%   50%   100%

The Green Revolution's Violence to Health

The challenge of allowing women access to food production is a global issue, but the US may display the ugliest metric on women and farming at only 7%.

The 2012 Census of Agriculture revealed that 92% of the 2 million US farmers are non-Hispanic, white men. In Iowa, the second largest producing agriculture state, 99.3% of farmers are white men. In 2012, the average farmer in America was 58.3 years old. Women own only 7% of US farmland and account for only 3% of sales.

Not only does the USDA discriminate against women, but also against young people and non-white men. The US agri-industry has justifiably earned a socially unacceptable label – racist.

Food costs leave 1 billion people in constant hunger and over 3 billion from obtaining good nutrition. Food costs rise with the cost of fossil resources, especially fuel and fertilizers. Food prices increase locally and regionally when climate chaos diminishes or destroys crops.

High oil prices raise shipping costs, which are often at least 20% of the product cost.[182] Food gets transported great distances, often 5,000 miles or more. Oil byproducts are a significant component of fertilizer, which contributes about 30% to food grain cost. Between 2001 and 2007, high oil prices added 40% to the cost of growing corn, wheat, and soybeans.

Farmer risk. Modern food crops require extensive and expensive inputs. If any of the many agri-inputs fails availability at just the right time, or weather insults interfere, the crop may fail partially or entirely. Global climate chaos puts farmers in serious jeopardy of crop loss from drought, temperature spikes, severe storms, floods, winds, wildfires and rising oceans. Pests, insects and weeds, add to farmer's risk.

Farmers commonly take loans to pay for their expensive agri-inputs, which puts many farmers in economic risk. The risk is rising since total US farm debt has nearly doubled, 46% since 2010. Total farm-sector indebtedness currently stands at $407 billion with a 2018 interest cost of $21.9 billion.[183] Financial risk diminishes mental health. The CDC found that farmers have a 30% higher risk of suicide than the next highest profession.[184] The rate of farmer suicide is higher than veterans returning from war.

Weather risk imposes a severe burden on farmers. One temperature spike or storm can destroy their crop. Hotter temperatures do not bode well for farmers. Each 1° C (2.5 °F) temperature rise above the optimum during the growing season cuts wheat, rice, and corn yields by 10%. A 3°C rise cuts yields by 50% and a 5° C rise destroys the crop. Severe drought pushed a dairy farmer in Eastern Australia to say:

> *It's gotten to the point where it's cheaper to shoot your cows than it is to feed them.*

Poison risk. Roughly 95% of pesticides are not absorbed by crops and pollute local ecosystems. Pesticide pollutants cause diseases of major organs, including the brain, heart, lungs, respiratory and vascular system.

Dust, pesticides and poisons make farming dangerous

Dust, fertilizer, pesticides and ag chemicals make farming and living in farm communities dangerous. Farmers, ranchers and agricultural managers are among the top five most dangerous jobs in the US. They suffer a fatality rate of 27 per 100,000 workers. Animal production ranked #1 in the US for injuries with 67 injuries per 100,000 workers.[185] Farm workers make the most dismal jobs list. Their jobs require intense physical labor and they work in harsh conditions with very low pay.

Farmer risk includes fatigue, since they must work long hours in all types of weather, perform heavy labor and work on and around heavy machinery. Farmers are regularly exposed to dust, pollens, agricultural chemicals, pesticides and other poisons. Available labor for food production will decline even further as young people leave rural areas for better jobs and lifestyles in cities.

India has horrible experience with farmer risk and debt as over 300,000 Indian farmers have committed suicide when they could not pay their debts.[186] Analysis indicates IMA does not work on the small plots many of India's farmers cultivate because the agri-inputs are too expensive. Farmer risk is driving an increase in US farmer suicides too.[187]

Harm to rural economies. IMA practices drive small farmers out of business. Farmers know the IMA motto well: "Get big or get out." Industrial methods work effectively for large farms, but modern methods have eliminated most small farmers. Farm Aid estimates that in the US, 330 famers a week leave their land for good.[188]

> *The Green Revolution focused on the big three – maize, rice and wheat – and did not adapt the big three to Africa, India, and many other countries with small farms.* — *Bill Gates*

IMA waste and pollution overwhelms rural communities. Towns must install expensive equipment to remove fertilizer by-products from public drinking water supplies, but most water systems cannot remove pesticides. More than 13 million US households rely on untreated. private wells for drinking water.[189]

Summary

IMA methods put farmers it extreme risk. IMA farming is a risky business as farmers force nature to do their will with costly heavy machinery, fertilizers, pesticides and poison exposure. Modern farming imposes these health and lifestyle risks on rural communities. The same methods that injure farmers severely harm farmlands and rural ecosystems. Waste and pollution degrade air and water and destroys natural ecosystems.

The next section explores healthier, sustainable solutions for food production with the Emerald Renaissance.

6. The Emerald Renaissance

The Emerald Renaissance stops the systemic extraction, consumption, waste and pollution imposed by intensive mechanical agriculture. Growers learn from nature and apply biological solutions with microcrops to resolve global hunger.

Abundance employs a circular biological architecture that recovers, recycles and reuses natural resources that are cheap or free, to produce a multiproduct crop. A single algae microcrop may produce healthier food for people, animals or plants, plus several additional coproducts.

Emerald Renaissance - Nature's biosolutions.

Circular - biocycles nutrients; mutiproduct model.
Zero emissions - Zero waste - Zero pollution.

Cultivate inoculant
Grow culture

Sunshine
or LEDs

Solar energy

Harvest 2x /week /all year
Protein production >50x
terrestrial crops
Weather independent

Waste stream nutrients:
• Air
• Water
• Biosolids
CCU: Carbon capture
and utilization

Nutrients:
• Recover
• Recycle
• Reuse

Multiproducts

Multiproducts:
• Food, feed, fibre
• Fertilizer, biofuels
• Bioplastics
• Biomaterials
• Nutraceuticals
• Medicines

Abundant Agriculture. Preserve natural resources with biomimicry.
Recover, recycle and reuse carbon and other nutrients.
Energy source: sunshine or LEDs for photon acquisition.

Freedom foods grow with minimal or no fossil resources, free from weather risk and produce 20% extra fresh water.

The Emerald Renaissance

A series of integrated biotechnology innovations combine to make the Emerald Renaissance architecture healthy and sustainable for many generations. Abundance methods biomimic nature where resources are recovered, recycled and repurposed into new bioproducts.

Abundance

Abundance takes its name from the inputs, which are plentiful and often free, surplus or cheap – sunshine, CO_2 and wastewater. Abundance methods biomimic nature's oldest and most reliable food production system at the base of the food chain. Microcrops are the fastest growing organisms, use low cost or free agri-inputs and transform sunlight into chemical energy very efficiently.

Abundance methods enable the production of freedom foods that grow powered by photosynthesis with free solar energy. Microcrop cultivation uses only modest electrical energy to move and mix water around a circular raceway or other container. Gravity flow and bubbles support mixing with minimal energy.

Cultivation energy may come from solar or other renewable sources. Abundance growers have no need for the costly and dangerous mechanical machinery commonly used for IMA food production. Microcrops harvest solar energy, water and CO_2 and the power of photosynthesis to produce a wide array of bioproducts.

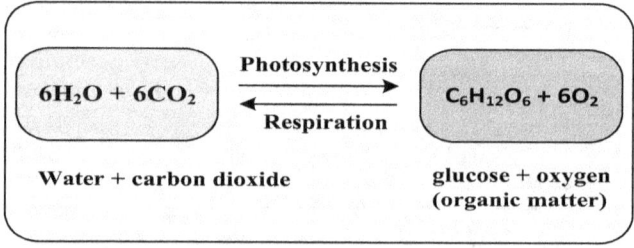

Biological production – solar energy and photosynthesis

With no human cultivation, today algae synthesize roughly 80 billion tons of organic matter daily, constituting about 40% of the total fresh organic matter grown on our planet.[190] Each day, algae supply 70% of the world's oxygen, far more than all the forests and fields combined. Fortunately, algae's many hungry consumers eat most the new biomass daily or the earth would be covered in algae.

Four agricultural methods

Abundant agriculture differs from three other forms on several dimensions, especially less consumption, waste and pollution, below. While IMA uses primarily mechanical solutions for cultivating crops, the other three methods use more biological solutions. In order to meet future food needs, all four forms of production will be needed.

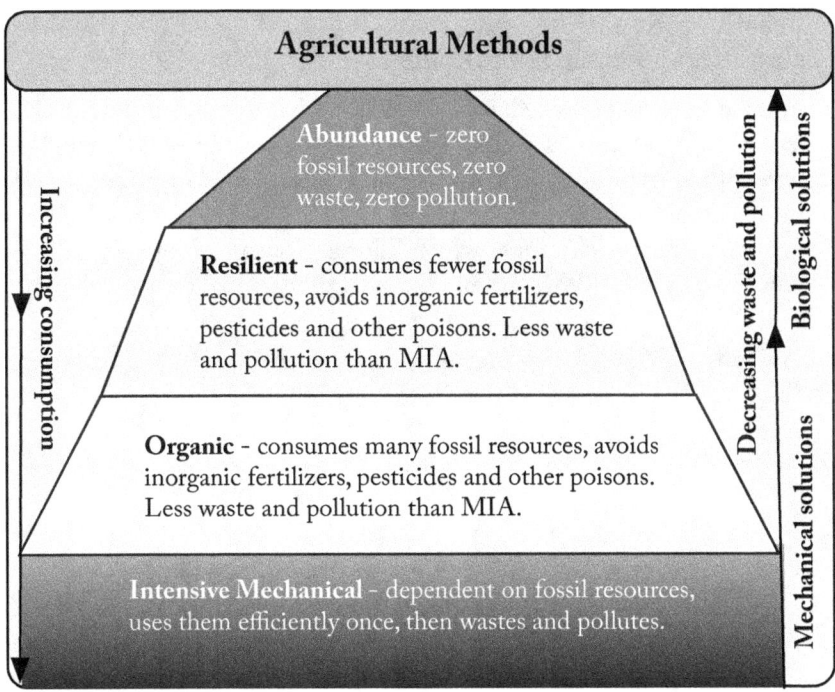

Agricultural Methods

Increasing consumption

Abundance - zero fossil resources, zero waste, zero pollution.

Resilient - consumes fewer fossil resources, avoids inorganic fertilizers, pesticides and other poisons. Less waste and pollution than MIA.

Organic - consumes many fossil resources, avoids inorganic fertilizers, pesticides and other poisons. Less waste and pollution than MIA.

Intensive Mechanical - dependent on fossil resources, uses them efficiently once, then wastes and pollutes.

Decreasing waste and pollution

Biological solutions

Mechanical solutions

Organic methods offer advantages over IMA in decreasing natural resource consumption, waste and pollution. Organic methods improve soil fertility with crop rotation, cover cropping, reduced tillage, and the use of compost. Tillage reductions avoids turning the soil and exposing it to air. Tillage avoidance improves soil organic carbon and nitrogen since they are not lost to the atmosphere.[191]

Resilient methods are targeted to smallholders to make crops more resistant to climate change. They use adaptive organic methods to improve systems, especially among the most vulnerable, strengthen natural resource management and reduce environmental degradation.[192]

Ten-zero food

Abundant agriculture applies natural biological solutions that biomimic nature's unique ability for ecological recovery, reuse and renewal. Abundance growers cultivate microcrops such as algae that offer substantial advantages over field crops.

Emerald Renaissance model applies inputs that are surplus and usually free or cheap – waste streams from air, water and biosolids. Sunlight powers photosynthesis, which allows the biosystem to recover, recycle and reuse waste stream nutrients. Microcrops grow more than 50 times faster than terrestrial crops, which allows growers to harvest daily or every few days.[193] In covered microfarms, growers harvest year-round.

Freedom Food - Ten Zero Food				
Zero fertile cropland	Zero fresh water	Zero fossil fuels	Zero chemical fertilizer	Zero pesticides and poisons
Zero waste	Zero pollution	Zero GMO	Zero hidden hunger	Zero empty calories

Freedom foods avoid the extraction, consumption, waste and pollution of precious fossil resources. Freedom foods biomimic nature, and she wastes nothing. She has produced superb and sustainable food for trillions of hungry consumers daily for over 3 billion years without consuming fossil resources. Freedom foods operate on nature's basic principle; **your waste serves as my food**. Freedom foods biocycle nutrients, which saves natural resources for future generations and substantially lowers growers' cost and health risk.

Ten-zero foods do not compete for resources with fossil foods because microcrops thrive with net-zero fossil resources.

The 10 zeros include:

1. **Zero carbon** emissions occur because every metric ton of fossil food consumes nearly two tons of CO_2, methane or other carbon source. Zero animals eliminate countless resource sinks and pollution. The only freedom food emission is pure oxygen.

2. **Fertile cropland** can be preserved for other uses as microcrops grow in raceways or containers. Microfarms may be sited on non-arable land, including vertical farms, deserts, mountains, cities, landfills, rooftops or oceans.

3. **Fresh water** can be saved to support families, households and industries. Microcrops flourish in nearly all kinds of water; waste, brine, brackish and highly saline ocean water. Microcrops create 20% more freshwater than they use as part of their multiproduct production suite.

4. **Fossil fuels** do not drive up grower costs or pollute the atmosphere because microcrops are produced using renewable energy-powered electric grids. Microcrops grow low on the food web and need only a tiny fraction of the energy required and consumed each growing season by fossil foods.

5. **Chemical fertilizers**, often 40% of an IMA farmers' production cost, are nearly zero for microfarmers who recover, recycle and reuse precious nutrients from waste streams. Freedom foods production biocycles nitrogen, phosphorus, potassium and other nutrients, which eliminates grower costs. Zero emissions eliminate fertilizer pollution to air, waterways and biotas.

6. **Pesticides, herbicides and fungicides** are malicious and nasty for people, producers and our planet. Microcrops grow without agri-chemicals and poisons, which eliminates pesticide residue on

73

produce and deadly farmer, family and community exposure. Clean production eliminates the ruinous poison invasion that maims or kills plants and aquatic life in rural environments.

7. Freedom foods create **zero waste** and no landfill because the only waste biomass, about 9% ash, becomes a nutrient source biocycled in the next culture. In contrast, about 60% of a steer on the hoof is non-edible waste. Some field crops like corn have even higher biomass waste.[194]

8. **Zero GMO monocultures** are needed because microcrops are extremely diverse, with over 100,000 times more species available than terrestrial crops. IMA monocultures degrade and destroy cropland, ecosystems and flora and fauna diversity. Freedom foods enhance biological diversity while producing good food and valuable bioproducts. The majestic diversity of microcrops ends many nature and political controversies over GMO foods.

9. **Zero hidden hunger** transfers from these foods to consumers because microcrop nutralence is an order of magnitude higher than terrestrial crops. Each tiny cell contains the entire nutrient package to support healthy growth and development.

10. Consumers will eat no **zero empty calories** because many are whole foods and do not require processing. Every bite contains 2 to 20 times more nutrients than fossil foods.

Freedom foods preserve natural resources and create zero air, water and ecosystem pollution. In sharp contrast to IMA methods, freedom food cultivation systemically cleans air, water and biosolid waste streams while producing a wide array of valuable bioproducts.

Four zero production

Within the next few years, our Emerald Renaissance team plans to integrate the ten zero food production with the Green New Deal's four zero clean, ecosmart production. Thomas Friedman notes:

> *Clean power, clean cars, clean manufacturing and efficient buildings make everything we want to achieve in our society easier.*[195]

The Emerald Renaissance

Hal Harvey, CEO of Energy Innovation, cites four zero carbon emissions.[196]

1. **A zero-carbon grid.** The electrical grid energy comes from solar or other renewables and integrates with smart storage systems.

These grids can become the cleanest and least-cost energy resource. Microcrops require very little energy beyond sunshine, which can be supplied easily by a zero-carbon grid.

2. **Zero-emission vehicles.** An electric vehicle charged on a zero-carbon grid allows zero-carbon transportation.

Freedom food production does not require large vehicles or heavy machinery, which makes electric vehicles efficient and cost effective.

3. **Zero-carbon/zero-energy buildings.** These buildings produce more energy than they consume.

Measured by the total cost of ownership, zero carbon buildings are cost competitive and usually cost less to own over the building's life. The Emerald Renaissance team plans to build zero-carbon and zero-energy buildings, made out of primarily renewable materials.

At stage 2, buildings such as for an eco-city and vertical farms will be made from biodegradable algae-based construction biomaterials. These are 100% renewable, extremely insulated, net energy positive and biodegradable. Food packaging material will also use microcrop-based fibers, bioplastics, and composites that capture and sequester massive amounts of carbon.

4. **Zero-waste manufacturing.** Waste increases carbon emissions and squanders energy, which tax both the producers' budget and our earth. Abundance methods avoid waste.

We plan to manufacture algae-based foods with 3-D printers that eliminate waste. Imagine a juicy microcrop hamburger that tastes great with zero fossil resource consumption or waste.

Microcrops offer the ideal media for 3-D printing. These nanocells flow smoothly through the printer cam and can be layered successively to produce foods of nearly any type, including pizza. A microcrop burger, compared with a beef burger, has 2.5 times the protein, 10 times the micronutrient density and healthy fats, including omega-3s.

The Impossible Foods burger has already proven the taste match using microcrops that evolved from algae – wheat, potato and soy protein. Burger ingredients include coconut oil and many added vitamins and nutrients.[197] Algae-based burgers do not have to add fossil oils and nutrients taken from other foods. Food scientists also have made burgers using vegetable protein and algae-cousin microcrops; yeast and fungi.

Freedom foods are made from beautiful and diverse microcrops.

Ten new freedoms

Freedom foods reinvent our food supply from the foundation of the food chain and supply 10 novel freedoms.

Freedom Food - Ten New Freedoms
Free consumers to make smart choices for foods that are healthier for people, producers and our planet.
1. Make informed choices for healthier food.
2. Have access to affordable and healthy food.
3. Avoid hidden hunger and empty calories.
4. Stay free of pesticide and argi-poison residuals.
5. Liberate mothers and their infants from nutrient deficiencies.
6. Free society from endemics - obesity, diabetes and CHD.
7. Free the atmosphere from heat-trapping C and other GHG.
8. Free producers and their families from severe health hazards.
9. Free waterways from over-consumption and agri-effluents.
10. Free ecosystems from degradation with bio-restoration.

Freedom foods deliver unique health freedoms for consumers, including freedom from unhealthy fat and cholesterol, hidden hunger, empty calories and micronutrient deficiencies. These foods are also free of contaminants, including dust, pollens, allergens and pesticide residue. The use of microcrop veggie-based meat and dairy products avoids all the human and animal health, resource consumption and environmental issues levied by animal foods.

Freedom foods capture and repurpose carbon, nitrogen, methane and other agri-GHG that contribute 40% to climate chaos. Freedom foods create zero waste and instead of polluting waterways, the biosystems biocycle nutrients, which are repurposed into clean bioproducts. Freedom foods can clean water as part of a multiproduct growth system.

Farm families are freed from the health jeopardy they face today; heavy machinery, heavy labor, fatigue, dust, agri-poisons and pollution. Freedom foods can clean polluted ecosystems, rebuild flora and fauna biodiversity and bring dead, abandoned soil back to life.

Microcrops include many types of microorganisms, including algae, yeast, fungi bacteria, archaea, protists, plankton, diatoms and many others that are incorporated in symbiotic communities. Most microcrops grow in moist environment without roots. Plants that flourish without roots deliver substantial advantages over terrestrial crops.

The freedom foods discussion here focuses on the foundation plant, an incredibly diverse, nutritionally rich, and productive food source – algae. Other microcrops offer similar health and ecological benefits.

Algae provide beautiful and nutritious microcrops

Healthier food

Freedom foods are naturally healthy and can treat and, in some cases, prevent obesity, diabetes, and other Western diet diseases. Not only are they low in fat and cholesterol, but they stop empty calories with high nutralence.

These foods are naturally biodiverse, which eliminates the need for GMO monocultures. Unlike fossil foods, freedom foods are clean, free of dust, allergens, chemical fertilizer, pesticide, herbicide and fungicide residue.

Chocolate cake provides a tasty case study.

Attribute	Industrial food Boxed chocolate cake mix	Freedom food Chocolate cake from algae flour*
High in saturated fats	Yes	No, 80% less
High in cholesterol	Yes	No, 90% less
Empty calories	Yes	No, 100% fewer
Pesticide residue	Yes	No, 100% less
Genetic engineering	Yes	No, 100% less
Preservatives	Yes	Minimal, 90% less
High protein	No	Yes, 100% higher
High macronutrients	No	Yes, 100% higher
High nutralence	No	Yes, 100% higher
Healthy omega-3s	No	Yes, 100% higher
Antioxidants	Few	Yes, 100% higher
Vitamins & minerals	Few	Yes, 100% higher
Trace elements	Few	Yes, 100% higher
Diminishes food cravings	No, increases food cravings	Yes, reduces nosh

Table 6.1. *Industrial Food versus Freedom Food Chocolate Cake*
* Note: Algae flour and oils are available at Whole Foods.[198]

Microcrop nutrition

Microcrops offer superior nutrition compared with terrestrial crops. Over one hundred times more animals eat algae than any other food because the biomass optimizes nutritional needs. Algae's tiny cell size, only 5 μ, make the plant an ideal nutrient delivery system for plants, animals and humans. Each tiny algae cell supplies the full set of essential nutrients needed for cellular metabolism.

Compared with field grains, algae offer 300 to 500% higher nutralence – nutrient density, quality, quantity, diversity, bioavailability, and bioactive compounds. The tiny yet high-nutralent cells are immediately absorbed and bioavailable to each consumer.

Freedom food bioproducts

Abundance methods can produce practically anything sourced today from land plants because land plants evolved from algae, below. Rooted plants inherited the superb colors, fragrances, pigments and nutrition from algae and other microcrops.

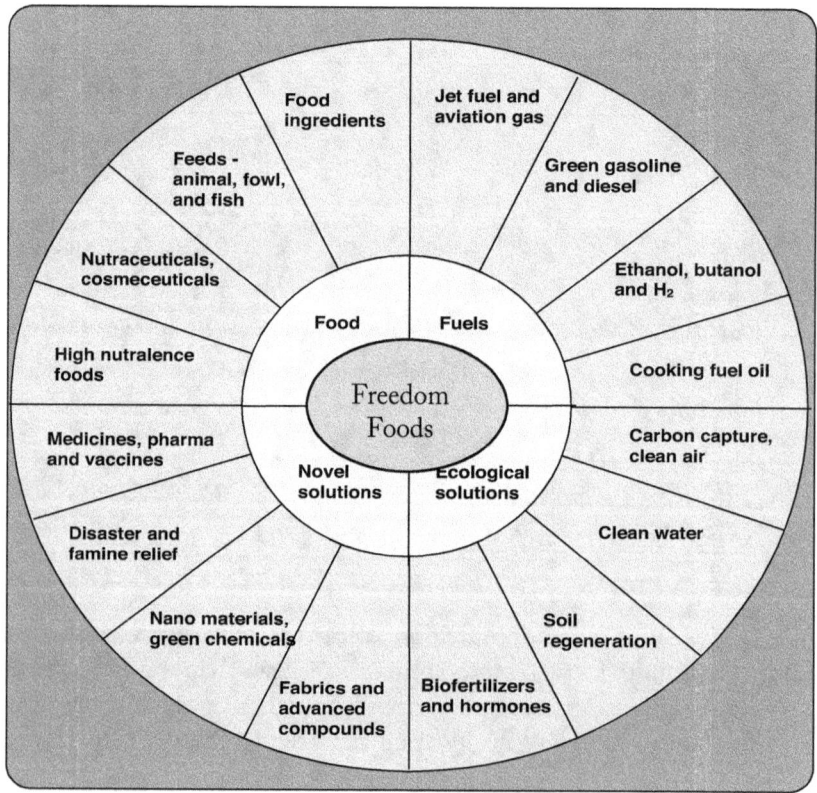

Fossil fuels, petroleum, coal, shale and natural gas are composed of algae or algae feeders fossilized from ancient oceans. Therefore, anything made from fossil fuels can be made from algae. Nature required 400 million years to produce fossil fuels. Algae can produce clean biofuels in a few weeks. Algae biofuels burn without black smoke particulates because they missed the hundreds of millions of years fossilizing carbon. Algae biodiesel burns cleanly, similar to vegetable oil. In fact, the tailpipe odor from biodiesel smells like French Fries.

Abundance growers can clean polluted water and air while creating carbon neutral and water neutral food and energy. Every ton of freedom food sequesters nearly two tons of CO_2. The only emission released by abundance methods is pure oxygen.

Freedom foods grown with abundance methods offer the opportunity to provide a food source that augments but does not compete for declining fossil resources required with industrial foods.[199] Scientific evidence and common-sense show that foods low on the food chain consume significantly fewer resources and are healthier for consumers.

Other sources expand on the many entrepreneurial opportunities with freedom food production in each sector.[200]

GMO

Freedom foods grow in such natural diversity that genetically modified organisms, (GMO) are not necessary. GMO crops may or may not bring health hazards. Several concerns are troubling:

- Many health impacts are likely to occur as a lagged effect, which could create serious health problems in the next decade.
- Transferring genes means the organism now contains new proteins, which can cause an unexpected allergic response.201
- GMO foods may increase risk of toxicity and chances of developing tumors, such as in the lungs, breasts or colon.202
- Modified genes migrate from the field into natural plants.
- GMOs substantially increase poison resistant weeds, worms and insects, which escalates the use of agri-chemicals and poisons.203

- Increases crop fresh water consumption significantly.
- GMO seeds are grown as monocrops that inflict brutal violence on cropland, water sources and rural communities. GMO agri-practice destroys natural biodiversity and accelerates extinctions.

The USDA, FDA, and EPA have failed to require labels on GMO foods. Lack of food labels combined with weak enforcement has enabled "GMO creep." In slightly over a decade, GMO foods have gone from zero to where today they make up a major portion of packaged foods.

Over 90% of U.S. food grains such as corn and soybeans grow in GMO monocultures that are refined into products loaded with fat, cholesterol, and calories but devoid of essential nutrients. The USDA has approved 81 GE crops, while failing to deny a single proposal.[204] Applications pending propose to use transgenics to alter up to 30 genes simultaneously for a single crop.

Micronutrients

Micronutrients create an Achilles' heel for IMA farmers. Each crop successively extracts more micronutrients from the soil, most of which are not replaced. Commercial fertilizers commonly bundle nitrogen, phosphorus and potassium, (NPK), for application. The result, nutrient dilution and hidden hunger, means that the produce may look fine but lack valuable micronutrients. Consumers, especially children, pregnant mothers and the elderly are extremely vulnerable to micronutrient malnutrition, which can have serious health impacts.

Abundance methods biocycle the full set of vital nutrients. Algae-based foods avoid hidden hunger and empty calories because algae grow to the limit of nutrients available in the culture. When the culture runs out of the first essential nutrient, algae cells rest and quit propagating. As soon as nutrients become available again, algae cells begin actively growing.

Weather resilience

A sustainable food system needs to produce foods that are resilient to weather and geography. Abundance growers cultivate food nearly independent of climate or weather. Algae have no true growing season, although most species hit maximum productivity during the normal growing season for terrestrial crops, from spring through autumn.

Cold weather species thrive in the winter, although many varietals slow or go dormant in extremely cold weather.

Cold climate growers may grow cold tolerant species in the winter. In Canada, Sweden, Iceland and other cold regions, growers add heat to accelerate biomass growth. Heat and light energy may come from renewable energy such as solar, wind, waves or geothermal.

Some species prosper in the cold, even thrive under the polar ice sheets. A 2018 study found marine algae plants flourish beneath the ice that covers the Greenland Sea.[205] These phytoplankton create the energy that fuels ocean ecosystems by providing energetic nutrition for the aquatic food chain. Algae produce about half of this energy under the sea ice in late winter and early spring, and the other half at the edge of the ice in spring. Many valuable aquatic creatures such as krill, shrimp and other shellfish depend on marine algae. Whales get their energy from eating algae directly or from eating the plentiful algae feeders such as krill.

Weather risk puts IMA farmers in serious jeopardy as a single weather insult can destroy an entire crop. Most IMA crops are extremely vulnerable to a slight, 2° C, temperature fluctuation during the growing season. Temperature insensitivity adds to farmer risk and severely limits available geography for farmers.

Robert Henrikson in his covered smart microfarm

The Emerald Renaissance

Peace microfarms

Freedom food growers cultivate algae and other microcrops in peace microfarms, above. These highly productive biosystems are called peace microfarms because they avoid consuming the precious natural resources that lead to war. Robert Henrikson calls these biosystems smart microfarms because they operate with smart sensors, monitors and control technologies that Robert developed. Robert Henrikson built excellent educational websites that share pictures and videos of microcrop production: Smart Microfarms and AlgaeCompetition.com.

Open microfarms are modestly sensitive to weather. An acre raceway that typically produces 150 kilos of biomass a day but may produce only 75 kilos during a storm or on cloudy days. When conditions are not ideal, algae simply rest. As soon as sufficient sunlight appears, production returns to normal. Since growers can harvest year-round, stormy and cloudy days have only modest impact on total production.

Open raceway and closed biosystem

Microfarms are flexible microcrop platforms that produce foods from plants lower on the food chain than modern industrial crops. Extensive research shows that foods low on the food chain, such as algae, provide superior nutrition for people, animals and plants than fossil foods with roots higher on the food chain.

Algae, and the other microorganisms algae attract, grow protein and other nutrients 50 times more productively than land plants. This means a corn farmer would have to farm one hectare for 50 years to produce as much protein as a freedom food grower cultivates annually on one hectare. Growers can double the biomass daily, which enables harvest of half the biomass every day the sun shines, year-round.

Direct sunlight may be too strong for algae. Most algae can use only about 0.1 the amount of light the cells receive from direct sunlight. Growers may use shade or diffuse sunshine to maximize production.

Covered or closed microfarms allow growers to produce algae food in practically any climate, altitude, latitude or geography. Growers in extreme locations may need to supplement with high-efficiency LED lights when sunshine is insufficient.

Microfarms in controlled environments such as a greenhouse with optional LED lights make production weather independent. Growers use sunlight when available and LED lights when solar energy is insufficient. Several vertical designs capture solar energy in the top culture and use LED lights to augment sunlight on lower layers. Some growers use highly efficient LED lights to extend biomass growth into the night and on cloudy days. Several companies are experimenting with 24/7 microcrop production with variable LED lights.

Please see AlgaeCompetition.com for imaginative microfarm graphics.

Visualize our Freedom Food Future – AlgaeCompetition.com

Local production

Food miles represent the distance food travels from seed to table. IMA foods often must travel a substantial distance from farm to processing center. Then, the food travels to regional distribution center, through the supply chain to retail stores and finally to the consumer's home, which may be thousands of miles.

Freedom foods thrive in nearly any geography without arable land.

Freedom foods may be grown locally in microfarms and eaten fresh, frozen or dried, without additional processing, stabilizers or preservatives. Freedom foods can be grown in cities, relieving urban areas of thousands of diesel trucks and their black soot particulates.

Slow growth, a full growing season – 90 to 120 days – forces IMA farmers to plant only a single crop a year. Every day after the seeds are in the ground, farmers risk losing their crop from weather, water, weeds, mildew, molds or pest vectors. Abundance growers cultivate their crops in microfarms and may harvest daily, 340 days a year.

Microcrop growers harvest more food, more often, improving their profits and substantially reducing production risk. Microfarmers take time off and need about two weeks a year for biosystem maintenance.

Low productivity requires IMA farmers to invest in vast tracks of fertile cropland for production. IMA practice has eliminated small farmers in the US and globally as the cost of agri-inputs are too high to make production economic, except for local farmers' markets. Microfarms are scalable and may be sited on inexpensive, non-arable land, which supports food justice.

Mechanical methods produce food but create a litany of disfunctions, especially negative environment impacts and risks to farmers.

Mechanical IMA farming equipment is costly and risky

Abundance employs biological solutions that biomimic nature and do no harm to the environment while providing safety to growers, farm animals and farm communities.

IMA's dependence on heavy equipment, inorganic fertilizers and agri-poisons combine to degrade soil as well as put farmers at risk of accidents and poisoning. Abundance methods eliminate the use of heavy equipment, (below) as well as chemical fertilizers and pesticides.

IMA production degrades and destroys cropland from systemic extraction of nutrients and humus from cropland. Abundance methods use biocycling to avoid waste and can restore health and vitality to ecosystems as they produce food and other valuable bioproducts.

Abundance methods create zero GHG emissions. Microfarms capture enormous amounts of harmful emissions – carbon, nitrogen oxides, methane and ozone – and emit only pure oxygen.

IMA creates wide-spread water pollution from eroded fertilizers, pesticides and other agri-chemicals. IMA runoff water is that fraction of irrigation water withdrawal that overspills the field. Overspill averages 46% of all water applied to fields.[206]

The Emerald Renaissance

Abundance growers can clean waste water while producing good food and other bioproducts. Growers are not dependent on freshwater since they can produce with plentiful sources, including waste, brine and even ocean water.

Summary

The Emerald Renaissance offers solutions for many of the fatal errors made in the Green Revolution and IMA practice. Most notably, the focus on biological solutions and biocycling ends the costly and wasteful IMA practice of systemic extraction, waste and pollution.

The next section examines the extraordinary freedom foods diet benefits.

7. Freedom Foods Diet – Healthy and Clean

Freedom foods offer numerous liberties to people, producers, and our planet while preserving natural resources.

Key takeaways

- What are abundance methods for sustainable food production?
- What are the health benefits delivered by freedom foods?

A sustainable food production system should provide foods that are healthier for consumers, producers and our planet. These foods should enhance the environment and provide positive sustainable social and economic benefits. Foods should prevent and treat malnutrition in all its forms, especially micronutrient deficiencies that cause stunting, anemia in reproductive age women and overweight.

Freedom foods allow consumers to optimize health at every life stage and deliver superior nutrition and taste without pollution or waste.

Freedom Foods Diet
Whole, High Nutralence Natural Foods
Healthier for People, Producers and our Planet

People	Producers	Planet
• Disease protection	• Low production risk	• Preserve resources
• No hidden hunger	• Multiproduct revenue	• Stop waste
• Full micronutrients	• No heavy machinery	• End pollution
• Stop + treat obesity	• Geo-independence	• Restore diversity
• Restore health	• Climate independent	• Restore ecosystems

Freedom Foods Diet – Healthy and Clean

Health and safety

Freedom foods free consumers to make healthier food choices. Freedom foods, compared with modern processed foods, deliver 2 to 10 times more nutrient density per bite. The nutrient set also provides 10 to 100 times more nutrient diversity and bioavailability. Freedom foods do not suffer from hidden hunger because microcrops stop growing if the culture lacks even a single essential nutrient.

Freedom foods contain many beneficial bioactive compounds, most of which are not found in rooted crops. Freedom foods halt the obesity and diabetes epidemic because these highly nutritious foods deliver very low fat, nearly no cholesterol and deliver healthy omega-3 fatty acids. Freedom foods can also protect against and in some cases provide therapeutic treatment people with Western diseases.

Abundance farmers create a positive eco-footprint, cleaning and improving the environment while producing food.

Social justice

After health, the most important goal for food production should be social justice, to assure everyone has access to affordable food.[207] Freedom foods can be grown locally in nearly any geography, including cities, urban and rural areas. Freedom farms distributed among both rural and urban communities can grow sufficient food to prevent malnutrition, nutrient deficiencies and stunting.

Food justice includes the ability of all people to have access not just to good food, but also the agri-inputs to produce the food. Freedom foods lower the barriers to food production. Biocycling nutrients substantially lowers grower costs. The avoidance of heavy labor, huge equipment, pesticides and other poisons allow nearly anyone to operate a microfarm. Several microfarm designs accommodate people with disabilities.

Food costs will go down with abundance methods. Local production alone can cut food costs by 50% by reducing or eliminating processing, packaging and transportation costs. Freedom food costs do not follow the rise in cost of fossil resources, because fossils are not used. [208]

Freedom Foods Diet – Healthy and Clean

Economic benefits to rural economies accrue by creating good jobs, producing desirable, nutritious food locally, and repairing local ecosystems. In some communities, the process of cleaning waste water may have more value than food production.

Freedom food production emits no unpleasant odors. The only emission from abundance methods is pure oxygen, which creates the pleasant smell similar to walking in the woods. Safeguards are necessary to assure the demise of waste stream parasites and pathogens. Freedom food growers use simple solar heaters, UV light or other technologies that has been used effectively for over 40 years.

Freedom to choose healthy food

Most contemporary consumers, even in the US, do not have access to healthy foods. The CDC published the Modified Retail Food Environment Index that shows that 9 out of 10 families lack access to retailers that sell healthy foods.[209] Freedom foods can change the access problem by providing local foods that are fresh and healthy because growers can produce these foods almost anywhere.

Many microcrop-based functional foods are available and provide excellent nutrition. The USDA spends billions of dollars on fossil food research, subsidies and crop insurance, but ignores natural foods, which get none of those substantial benefits. Less than 1% of the USDA R&D budget supports organic production. The US government has made small investments in algae research recently, but the agri-policy focus has been biofuels with zero investment for sustainable and affordable freedom food.

Water

Freedom foods thrive without freshwater. Cultivation can use non-potable water sources including brine, brackish, agricultural, municipal or industrial wastewater, or ocean water. These water sources typically contain too much salinity or other pollutants for terrestrial crops.

Salt invasion from irrigation salt, fertilizers and agri-chemicals has destroyed millions of cropland hectares and many civilizations.

Terrestrial crops die in the presence of salt due to a plumbing problem. Salt ions in solution are too large for root absorption. The ions clog up the roots and block water flow, which causes the plant to starve.

Algae evolved in ancient oceans where the plant had to adapt to high-saline conditions. Algae have no roots and absorb nutrients directly into the cell, independent of the in-situ saline concentration. Most species are already or can be adapted to saline water. Both macroalgae, (seaweed or sea vegetables) and microalgae thrive in salty estuaries and oceans.

Brine water makes up over 50% of the groundwater stored on the planet. Brine has a salinity of 3.5% to 26%. Brine water was produced from the drying of ancient oceans and salt lakes. Today, many industrial processes produce waste brine water, including food processing and desalinization.

Freedom food production in a raceway and PBRs

Brine water alone provides a sustainable freedom foods cultivation medium. When crude oil pumps remove oil from the ground, they often recover 20 barrels of brine water for every barrel of oil. Today, this nutrient rich water is pumped back into the well to create pressure for the extraction of additional crude oil.

Smart algapreneurs will cultivate freedom foods to remove the nutrients before the brine water is returned to the well. In many areas in Africa, brine water lies only a few feet under the ground and can be recovered by foot pumps.

Wastewater creates a huge cost for cities, industries and farmers. Freedom farms can transform those costs to positive revenue, creating valuable biomass while cleaning the wastewater. The oldest algae application in the US is wastewater treatment.

Oceans have served as the best natural recycler on our planet for eons. Marine ecosystems accept runoff sewage and botanical waste and recycles it into nutrients. Natural systems scrub toxins from the water and turn waste CO_2 into food and oxygen.

Freedom food producers biomimics the ocean's superb recycling process. Growers use a suite of natural methods including UV light, solar heaters or other tools to remove bacteria and pathogens.

Freedom foods grow to the limit of nutrients. If insufficient water or nutrients are available, algae take a rest and simply enter dormancy. When good growing conditions return, algae resume normal growth.

Sea Vegetable – Edible Seaweeds

Waste

Industrial agriculture produces massive waste and pollution. Peace microfarms can recover and repurpose bioenergy and nutrient waste into energetic foods. Abundance growers can clean polluted air, water and soil while they produce valuable algae biomass.

ZooPoo case study

Currently, waste streams represent a significant cost for farmers and for zoos. Zoos often have to pay more to dispose of their ZooPoo than they pay for animal feed. ZooPoo includes animal, botanical and trash wastes. Zoos must pay substantial costs for handling, storing, inspecting, hauling and disposing of ZooPoo. Currently, ZooPoo adds tons of biomass to landfills, which degrade the environment.

The amazing thing about ZooPoo is its lost value. Animal manure and botanical wastes retain roughly 60% of the bioenergy that was originally stored in the plant. Even more importantly from a farmer's perspective, ZooPoo retains 80 to 90% of the original plant nutrients. Elephants are the poster child for ZooPoo since elephants leave 95% of the original botanical nutrients in their poo. Elephants eat heavy roughage.

Elephant poo retains 95% of the original nutrients

Abundance methods can recover and reuse the zoo waste stream and transform this huge cost to a profit center. The combined value of the reclaimed bioenergy and nutrients could cover the entire cost of animal care. A smart zoo will create an additional ZooPoo revenue stream with a world-class ecotourism exhibit that conveys how abundance methods support sustainable green living.

Toxic waste

Human, industrial and agricultural waste streams often contain heavy loads of fertilizers, pesticides, herbicides and fungicides that make them unacceptable for use on field crops. Tiny algae cells absorb individual elements from complex compounds such as pesticides, which detoxifies the agricultural poison. Current technology offers several algae solutions that detoxify waste streams.

Drugs create a serious problem with manure as fertilizer. Organic food producers want to use animal manure as fertilizer but in many cases they cannot due to the pharmaceutical drug problem. Over 80% of all antibiotics produced in the US, roughly 25 million pounds a year, are fed to meat and dairy animals.[210]

Feeding pharmaceuticals to animals sustains a growing demand for meat but it generates public health fears from the expanding presence of antibiotics in the food chain. About 90% of the drugs in animals and humans end up being excreted either as urine or manure. Food crops absorb and concentrate antibiotics and other drugs when grown in soil fertilized with livestock manure.

Widespread use of antibiotics leads to the growth of drug-resistant superbugs, which the CDC estimates now kills over 23,000 people in the US each year.[211] A recent study of 48 US pediatric hospitals found that drug-resistant infections in children had increased 700% in eight years, which the authors called "ominous."[212]

Heavy metals

Algae are voracious pollutant scavengers for a broad category of chemicals released into the environment from the domestic, industrial and agricultural sectors. Besides the usual organic and inorganic fertilizer residue compounds present in the wastewater, algae cells can also assimilate and/or break down more persisting molecules such as hydrocarbons, antibiotics, PPCPs, EDCs and heavy metals.[213]

Australia's James Cook University demonstrated algae are effective at bioremediation of CO_2 and heavy metals, (Al, As, Cd, Cr, Cu, Ni, and Zn), *in situ* at a coal-fired power station.[214] Macroalgae inoculates were grown in vertical tubes, (right) and moved to shallow wastewater ponds containing CO_2 and fly ash. The algae removed nearly all the metals.

Bioremediation of excess nutrients in wastewater by microalgae has been prevalent in the US for the past 70 years.[215] When algae bioaccumulates toxic wastes, the biomass may not be useful for normal bioproducts. The toxic biomass can be converted to biochar through pyrolysis or gasification. The biochar can be used as soil amendment with reduced risk of leaching of toxic material such as heavy metals, since the pyrolysis process integrates and binds up the metals in the solid matrix.

Bioremediation uses naturally occurring organisms as a treatment to break down hazardous substances such as waste or pollutants into less toxic or non-toxic substances. Bioremediation with nutrient recovery serves as the first step in nutrient cycling. Considerable bioremediation research focuses on the capture of nutrients harmful to the environment, animals or people, such as toxic heavy metals from pesticides.

The role of microorganisms in biotransformation of heavy metals into nontoxic forms is well-documented.[216] Understanding the molecular mechanism of metal accumulation has numerous biotechnological implications for bioremediation of metal-contaminated sites. Cell wall components of various algae groups is closely related to the metal binding capacity of algae. These biotechnology layers are described in *Adsorption and Absorption of Heavy Metals by Microalgae*, by Li Li.[217]

After bioremediation, algae have accumulated the toxic heavy metals so the biomass is not fit for anaerobic digestion or biofertilizer. Pyrolysis of the cultivated algae immobilized the accumulated metals in a recalcitrant C-rich biochar. While the algae biochar has 10 to 50 times higher metal concentrations than the algae feedstock, the biochar had very low leachable metals.[218] The metals were bound up (chelated) into the biochar matrix, providing nutrients for crops but sequestering the toxic metals for decades.

Bioremediation lays the foundation for nutrient cycling. Nutrient capture in algae biosystems provides the feedstock for a wide variety of bioproducts. Since modern environments offer so many pollutive point sources for carbon and other nutrients, algapreneurs have many choices for nutrient collection. Carbon capture using current mechanical technologies cost \$90 to \$150 a ton, but those methods are not sustainable. Algae-based CO_2 capture and sequestration costs more than conventional methods currently but are more sustainable. In addition, algae produce bioproducts that can be monetized. Algae-based CO_2 and nutrient capture will present a profitable business in the near future with rising oil prices, carbon trading and social policies directed at polluters.

Algae's three immediate bioremediation contributions to agriculture will be recovering phosphorus, reducing nitrogen and carbon wastes, and capturing and sequestering toxic heavy metals from pesticide residues. The strongest leverage to agriculture from algae bioremediation may be algae biofertilizers.

Most importantly, algae-based CO_2 abatement enables monetizing carbon credit + nutrient value + biomass value for bioproducts. A coal power plant produces about 1 ton of CO_2 for every MWh of energy produced. Yield of algae biomass per hectare is about 0.3 to 1 ton per day. Algae biotechnology will offer a safe and sustainable solution to the problems associated with CO_2 emissions from coal power plants and other carbon sources.

Air pollution

Abundant agriculture provides significant air pollution solutions. Power, cement and manufacturing plants produce the heaviest loads of carbon and other pollutants to the atmosphere. Several algae producers such as Carbon Capture Corporation are designing systems that sequester carbon from exhaust plumes.[219] Current technology allows microfarms to capture only part of the waste stream because power plants operate 24/7 while algae grow during daylight hours. Since every ton of algae captures two tons of CO_2, waste plumes provide substantial economic benefit.

These pollution sources also emit tons of black carbon particulates, nitric oxides and sulfur that cause a myriad of harmful effects on the respiratory and cardiovascular system, the heart, blood, and blood vessels. Black carbon exposure has been linked to over 300,000 premature deaths a year in the US.[220] Globally, black carbon pollution causes 3 million premature deaths annually and over 74 million years of healthy life lost.[221]

Microfarms can capture black carbon particulates and recycle them into valuable bioproducts. Algae assimilate the particulates in a manner similar to the assimilation of CO_2 and CH_4, (methane) which provides the carbon that algae combine with water to make simple sugars.

The following tables summarizes the substantial differences.

Attribute	Green Revolution	Emerald Renaissance
Method	Intensive mechanical	Abundance
Production	Monoculture	Diverse cultures
Resource model	Linear – single use	Circular – biocycle
Climate risk	Intolerant, high risk	Independent
Fossil resources	Dependent	Independent
Sustainable	No	Yes, many generations
Production cost	High	Low
Cycle nutrients	No, one-time use	Yes, > 10x use
Produces	Single commodity	Multiple bioproducts
Input costs	High and increasing	Low and will stay low
Time to harvest	Full season – 120 days	Every few days, all year
Biodiversity	No, monocultures	Highly biodiverse
Crisis defined	Insufficient protein	Insufficient nutrients
Reward system	Tonnage, yield weight	Nutrition and yield
Healthy food	No, empty calories	Yes, highly nutritious
Nutrition	Poor, low density	Excellent, high density
Obesity	Causes obesity	Moderates obesity

8. Freedom Foods Preserve Fossil Resources

Abundance methods grow healthier and sustainable food free from fossil resources, waste or pollution. Growers use cheap and often free resources, saving agri-input costs and preserve precious fossil resources for future generations.

Key takeaways

- How do freedom foods preserve natural resources?
- What advantages do freedom foods deliver to consumers?
- What advantages to microcrops provide to growers?

Freedom foods provide splendid natural foods similar to nature and without waste or pollution. Nature wastes nothing as the death of one organism provides the vital nutrients that give life to the next. Abundance methods do not compete with IMA because growers use bountiful resources recovered from waste streams.

The IMA linear model forces a one-time use of all agri-inputs puts farmers at extreme financial risk from crop failure, (below). IMA creates excessive costs, labor, waste and pollution. The circular abundance method biocycles nutrients that saves farmers money, time and labor. Biocycling eliminates wasteful emissions.

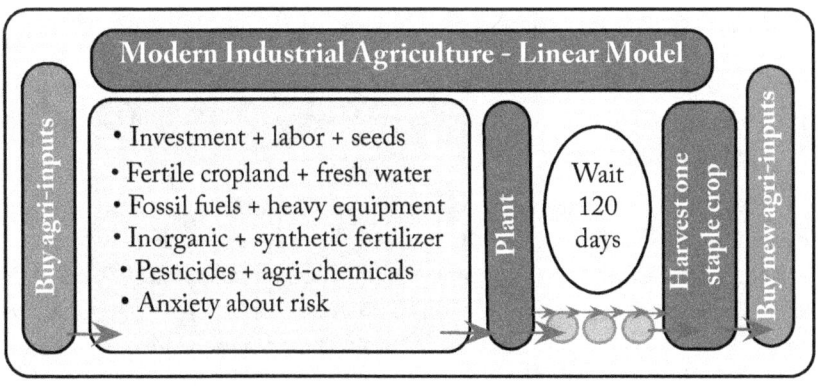

Abundance methods avoid the use of fossil resources, Table 8.1, by efficient carbon and nutrient biocyling to produce a wide array of hydrocarbon bioproducts. Abundance agricultural methods free growers from the substantial cost and labor of fossil resource extraction, consumption, waste and pollution.

Primary fossil resources		Micronutrients	
1. Fertile soil		12. Oxygen	(O)
2. Freshwater		13. Hydrogen	(H)
3. Fossil fuels		14. Sulfur	(S)
4. Inorganic fertilizers		15. Magnesium	(Mg)
5. Pesticides / herbicides		16. Boron	(B)
6. GMO seeds		17. Copper	(Cu)
Macronutrients		18. Chlorine	(Cl)
7. Nitrogen	(N)	19. Iron	(Fe)
8. Phosphorus	(P)	20. Molybdenum	(Mo)
9. Potassium	(K)	21. Manganese	(Mn)
10. Calcium	(Ca)	22. Nickel	(Ni)
11. Carbon	(C)	23. Zinc	(Zn)

Table 8.1. Fossil resources required for intensive mechanical agriculture

Freedom foods grow in peace microfarms that recover, recycle and reuse carbon and other nutrients, reducing farmer costs and avoiding nutrient scarcity. Many experts predict war over the obvious resources; fertile cropland, fresh water and fuels. Careful analysis shows wars may be fought over macronutrients such as phosphorus and micronutrients such as copper and zinc, which are in very limited supply.

For example, failing sufficient zinc in the soil, terrestrial crops fail to fruit, (grow seeds), which eliminates not just the food portion of the crop but seeds for subsequent years.

Organic production methods recycle compost and avoid some chemical fertilizers and pesticides, but remain heavily dependent on fossil resources.[222] Fewer than 1% of US farms are organic, largely because large monocrop farms have not found an economically feasible means to transition.[223] Organic farming requires free range animals, which consumes considerable land. Organic farms have trouble sourcing sufficient organic fertilizer. Compositing requires extensive labor and energy gathering, piling and then applying compost that must sit for a year, while it diminishes in mass by a factor of 10. Much of the nitrogen volatizes during composting, which reduces nutrient value and adds harmful nitric oxides to the atmosphere and waterways.

Regenerative agriculture offers a positive future for small farmers and addresses many of the flaws inherent in IMA. Regenerative methods avoid pesticides, synthetic fertilizers and monocrops with a strategy to regenerate topsoil, increase biodiversity, and many other benefits. Regenerative farming takes a biological approach to enhance soil structure, increase yields and decrease external agri-inputs.[224]

Regenerative agriculture reduces the consumption of fossil resources and pollution significantly. The GRAIN Report provides a thoughtful road map of how small farmers can feed the world using regenerative agriculture.[225] Regeneration International offers another good report on how small famers can reverse climate change.[226]

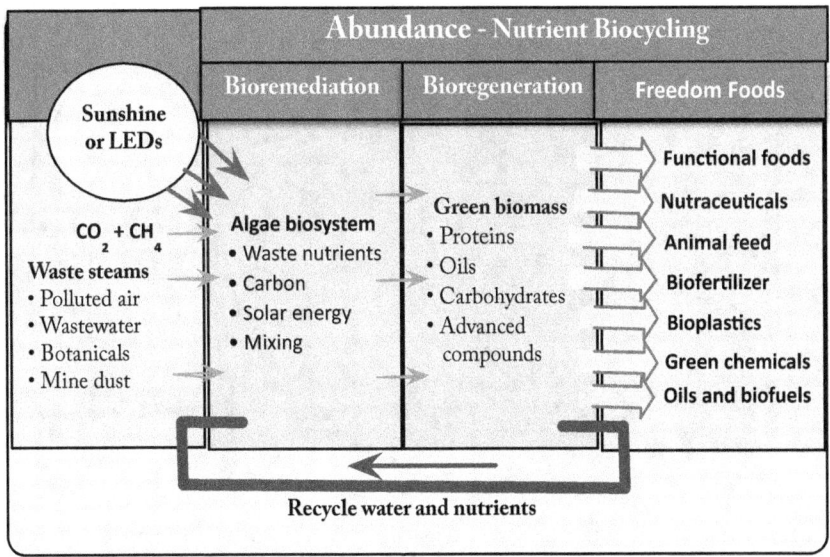

Abundance methods, above, are regenerative and assure sustainable food production for many generations because most the resources are recovered, recycled and repurposed into new bioproducts.

Freedom Foods Preserve Fossil Resources

The Emerald Renaissance offers key advantages with the use of biologically smart abundance methods and freedom foods.

Natural resource preservation

Terrestrial crops require massive tracks of relatively flat fertile soil. Abundance production requires none because the nature the product, algae, grows in raceways or containers that can be placed on non-arable land such as deserts, barren land, mountains, rooftops or oceans.

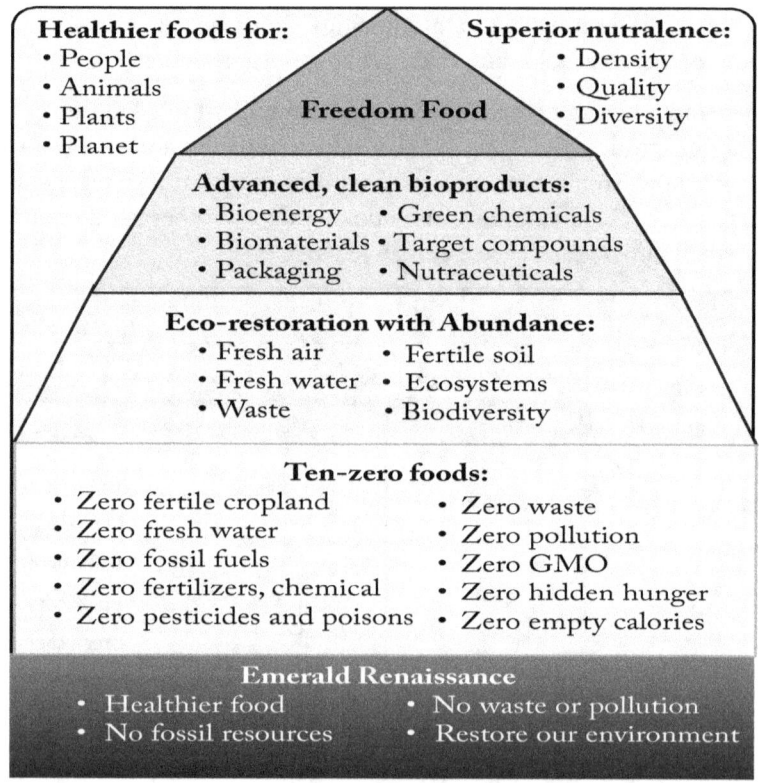

Freedom foods, above, save fresh water for other purposes. Algae production flourishes in brine, municipal and industrial waste, irrigation withdrawal and ocean water. Several freedom food models create fresh water as a coproduct of production.

Freedom Foods Preserve Fossil Resources

Terrestrial crops use diesel fuel intensively, often 10 to 15 liters per ton of production. Farm machinery uses diesel for cultivation, planting, fertilizing and agri-chemical applications, as do the trucks used to transport biomass to processing plants. Microalgae require no heavy farming equipment and no transport of biomass to distant processing plants. Freedom foods do require modest energy for cultivation and harvesting, which can be supplied by solar or other renewable sources.

IMA farmers apply millions of tons of chemical fertilizers on fields every year with tractors and heavy equipment. Extraction, transportation and processing chemical fertilizers add substantial GHG to the atmosphere.

Manufacturing one ton of nitrogen fertilizer, anhydrous ammonia, consumes 33,500 cubic feet of natural gas.[227] Corn production consumes a ton of nitrogen fertilizer for 13.3 acres, which creates a carbon footprint equivalent of driving a car 8,7000 miles. Much of the nitrogen applied to the field volatizes into dangerous nitrogen oxides, which add to smog, air pollution and more GHG.

Terrestrial crops, especially GMO varieties, require substantial amounts of pesticides that kill invading insects, herbicides to protect from invasive weeds, and fungicides that kill fungal vectors. GMO crops maximize production by subtracting energy from roots and by tight planting. Both actions increase the need for more water as short roots cannot reach deep for moisture. GMO crops also require more agri-poisons because they are more vulnerable to invasive vectors such as resistant weeds and insects. Less than 5% of these poisons enter the crop, while 95% erodes from the field poisoning the local ecology.

Freedom food production uses no pesticides, which avoids the substantial cost, waste and poisoning of surrounding eco-communities.

Multiproduct production

IMA farmers produce a single monocrop commodity. The staple crop makes farm revenue dependent on the commodity price, over which they have no control.

Freedom Foods Preserve Fossil Resources

Freedom food growers avoid commodity crops because multiproduct production creates substantially higher revenue. While some of their co-products may be commodities, such as biofertilizer or animal feed, their primary product might be a nutraceutical such as astaxanthin that creates more revenue than all the coproducts combined. Algae biomass includes proteins, lipids, carbohydrates and specialty compounds, below.

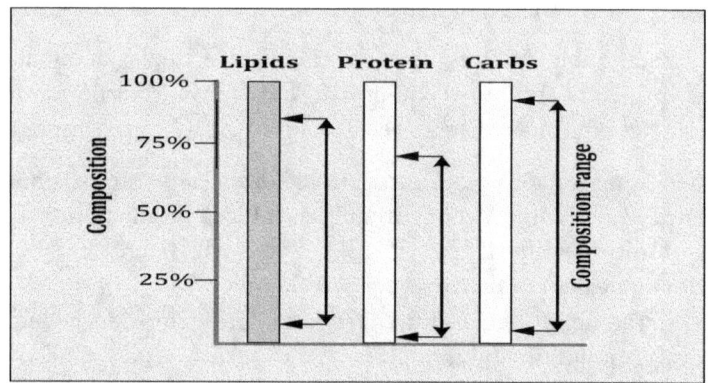

Proteins are large organic compounds made of amino acids arranged in a linear chain connected by peptide bonds. The plant's genetic code determines the sequence of the amino acids, but nutrient limitations may cause changes to the production of amino acids. Most proteins are enzymes that catalyze biochemical reactions and plant metabolism. Other proteins maintain cell shape and provide signaling functions.

Lipids are long carbon chain molecules. Lipids store energy for the plant and serve as the structural components of cell membranes. Starches are complex carbohydrates which are insoluble in water. Plants use starches to store glucose, plant sugar.

Specialty compounds include nutraceuticals, functional food ingredients, pharmaceuticals and medicines. The composition variation among species varies tremendously. Some algae hold 80% lipids while others are 66% protein and still others are 92% carbohydrates.

Freedom food growers use multiproduct production, below, which provides extraordinary value. Freedom food crops may grow continuously or in batch cultures, depending on the production strategy. In both methods, the rich biomass contains proteins, lipids (oils), carbohydrates, specialty target compounds and ash.

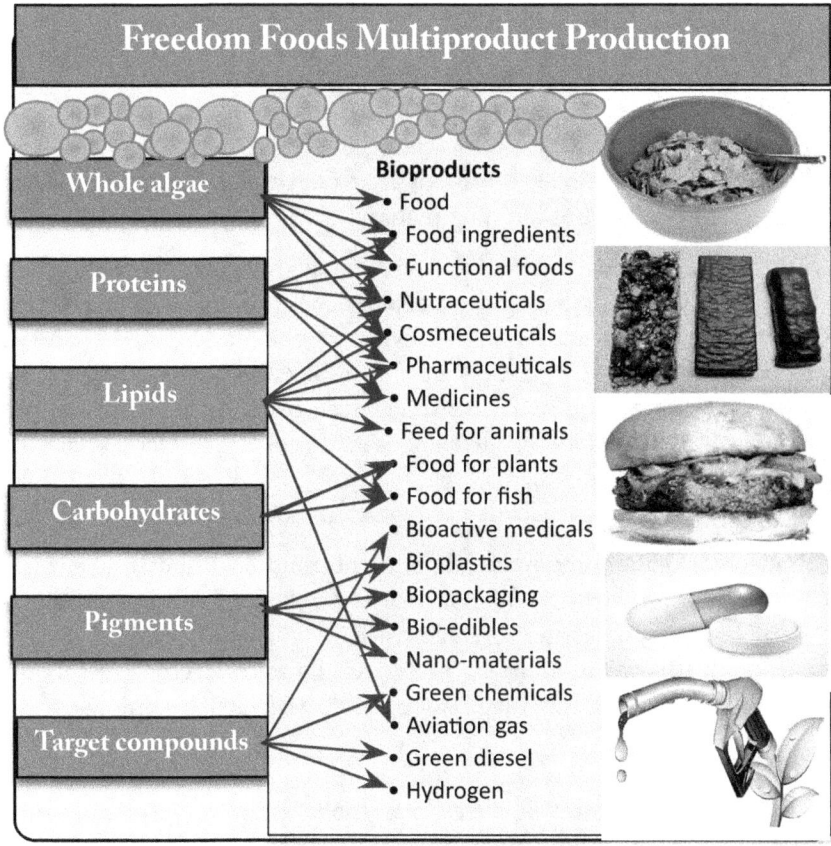

Growers select from among many diverse species to optimize the value of their multiple bioproducts. Growers also know that changing growing conditions can alter the biomass composition to overproduce desired compounds.

When algae are nutrient limited, such as nitrogen, phosphorous or sulfur, they decrease the amounts of essential polyunsaturated fatty acids produced and may yield lower quality protein with fewer amino acids.

Nutrient deprivation may cause algae to increase lipid production but unfortunately, nutrient deprivation often slows or halts propagation and growth. Bioengineers are searching for algae strains that increase lipids without nutrient deprivation.

How is bioplastic a food?

Conventional thinking would challenge the concept that freedom foods would include bioplastics, packaging, building materials and biofuels. Biofeed and biofertilizer are foods that deliver superior nutrition for animals and plants, but building materials?

All freedom foods bioproducts are biodegradable, plus some are bioedible. Microproduct biodegradable means products can be recycled immediately or degrade left in the environment in under 100 days. Algae biopackaging and bioplastic can make bioedible bags and serving utensils. After eating lunch, consumers can eat their fork and the bag that carried their food. While some would probably prefer to feed the biowaste to their pet llama, the materials are digestible by people. Imagine the value of packaging dog food in bioedible bags.

Biodegradable building materials can withstand environmental stresses during their product life. Some use special covering. When a building's useful life ends, the materials can be ground up and returned to an algae biosystem. Biocycling building materials allows microcrops to support life support and building materials on a 100-year Starship mission.[228]

Summary

Abundance methods and freedom foods preserve precious fossil resources for our future generations. The avoidance of fossil resources frees growers and ecosystems of waste and pollution. Freedom foods also free growers from the substantial costs for the annual agri-inputs that are used only once. Multiproduct production frees growers from the tyranny of a single commodity price for their monocrop. The ability to make multiple valuable products in the same microfarm gives growers nearly unlimited opportunity to mix and match bioproducts to the needs of their family and community.

9. What are Microcrop Advantages?

Microcrops significant competitive advantages challenge whether a terrestrial crop strategy makes sense today. Microcrops grow 50 times faster than rooted crops, grow with no fossil resources and create neither waste nor pollution. Each bite of the resulting biomass offers 2 to 10 times more nutrition than rooted staple crops.

Key takeaways

- How are microcrops different from field crops with roots?
- What advantages to microcrops deliver to growers?

IMA produces large terrestrial crops that are severely constrained by many physical features such as roots, stems, leaves and sexual elements that waste the plant's limited energy available for food production. In 1950, little was known about microcrops, which made land-crops the only option. Key tools that underlie the Emerald Renaissance did not yet exist, such as scanning electron microscopes and biotechnology.

How can microcrops produce food 50 times faster than staple crops?

The question for food production is not photosynthetic efficiency, (how efficiently photons are captured) but does the plant distribute its energy to grow food or non-food features?

Microalgae cells measure only about 5 microns across, (0.00004 inches). They are similar in size to a human red blood cell, or about 1/15th the diameter of a human hair. Algae display beautiful colors, textures and shapes when viewed under a microscope. Some microalgae aggregate in colonies that form chains or groups. Spirulina create colonies of single tiny cells that string together and are visible to the eye. Many macroalgae, seaweed or sea vegetables, also string cells together tightly.

What are Microcrop Advantages?

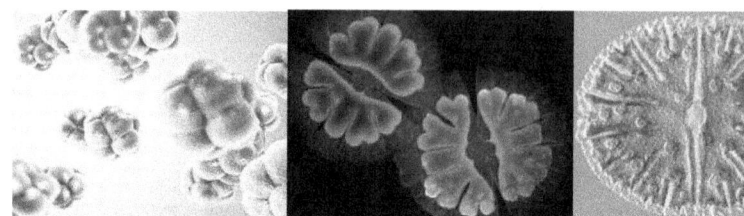

Algae nano-cells under a microscope

Even though algae are tiny, each day algae produce 70% of the world's oxygen, far more than all the forests and fields combined. Algae grow so fast that each day they produce 40% of the new biomass.

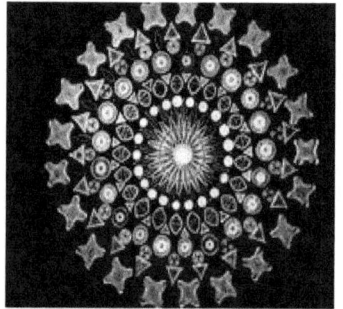

Algae are beautiful. British biologist Klaus Kemp is one of the last practitioners of the Victorian art of diatom arrangement, (at left).[229] These single-celled organisms are found in oceans all over the world. There are estimated to be diatom 100,000 species. They play a crucial role as one of the primary food sources for marine organisms, including fish, mollusks and tunicates, such as sea squirts. The New York Times hosts a beautiful video on diatoms.[230]

Phycology

Phycology, the study of algae, includes the study of prokaryotic forms known as blue-green algae or cyanobacteria.[231] Terrestrial algae live in or on soils and others live in symbiosis with lichens, corals and sponges.[232] The basic single-celled organism, algae, has the general appearance illustrated below.[233] The University of Montreal, U.C. Berkeley, University of Texas and others host culture collections of algae samples for sale and offer descriptive details and pictures.[234]

Eukaryotic green algae, (Greek for "true nut") plants have cells with their genetic material organized in organelles.[235] They create discrete structures with specific functions and have a double membrane-bound nucleus or nuclei. The prokaryotic cells of blue-green algae,

What are Microcrop Advantages?

cyanobacteria, contain no nucleus, cell walls or other membrane-bound organelles.[236]

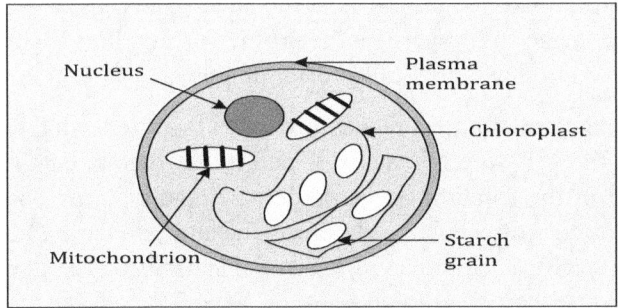

Alga cell – about 5 microns across

The first alga (singular) cell was among the earliest life-forms on Earth, probably about 3.7 billion years ago in oceanic environment synthesized by abiotic, high energy processes including lightning, ultraviolet radiation and pressure shock. The atmosphere was anaerobic with high levels of methane, hydrogen and ammonia, but no oxygen.

Algae are distinguished from the higher plants by a lack of true roots, stems or leaves. Some seaweed such as kelp (right) appear to have leaves or trunk. These are pseudo leaves made up of the same simple cell structure as the rest of the plant.

Many species are microscopic single-celled plants, including phytoplankton, diatoms and other microalgae. Macroalgae or seaweed are multicellular and may grow up to 90 meters such as kelp. Sargassum, a brown alga, has the distinct advantage of growing at one end and dying at the other. Sargassum can float on ocean currents for years.

What are Microcrop Advantages?

Major steps in cell complexity occurred with the evolutionary progression from a virus to bacterium and then from the prokaryotic cells of bacteria that had no cell walls, to the eukaryotic cells of algae. Cell walls enable algae to protect itself from the surrounding environment, typically water and pressure, called osmotic pressure.

Cell walls regulate osmotic pressure produced by water trying to flow in or out of the cell through its semi-permeable membranes due to a differential in the solution concentrations. Algae typically possess cell walls constructed of cellulose, glycoproteins and polysaccharides while some species have a cell wall composed of silicic (silicon) or alginic acid.

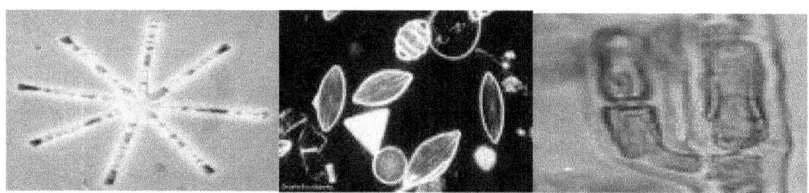

Algae break the rules for plant classification because they evolved in many different forms – single cells, multicellular plants, bacteria and in nearly infinite combinations. While the various species share certain characteristics, different algae display extraordinary variety in shape, size, structure, composition and color.

Single-cell growth advantage

Microcrops such as algae are differentiated from higher level plants because they generally:

- Avoid the substantial limitations imposed by roots.
- Are aquatic microscopic plants with chlorophyll-a and a single-cell body not differentiated into roots, stems or leaves.
- Display the ability to perform photosynthesis with the production of molecular oxygen, which is associated with the presence of chlorophyll a, b or c.
- Do not have specialized transport tissues or organs consisting of interconnected cells that move nutrients and metabolites among different sites within the organism.

What are Microcrop Advantages?

- Reproduce sexually or asexually to produce gametes that generally are not surrounded by protective multicellular parental tissue.237
- Are tiny cellular biofactories that use the substantial power of solar-driven photosynthesis for biomass production.

Cellular biofactories

Microcrop cells are tiny, 5 microns across, but efficiently manage millions of simultaneous photosynthetic operations, below.

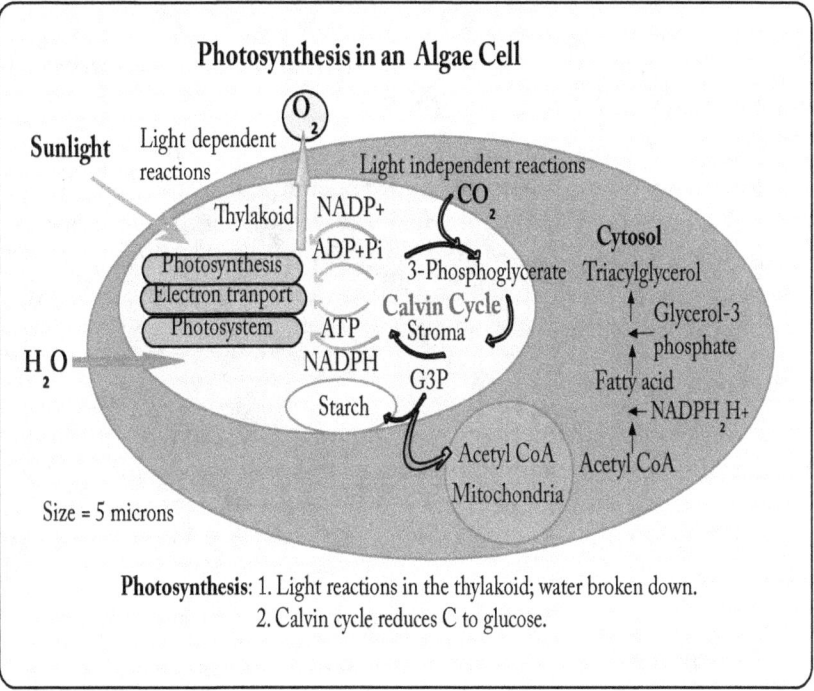

Photosynthesis in an Algae Cell

Photosynthesis: 1. Light reactions in the thylakoid; water broken down.
2. Calvin cycle reduces C to glucose.

Roots imposed significant constraints on land plants. Roots act like an anchor, tying the plant to the ground which makes the plant dependent on water and nutrients in situ. Land plants must grow and sustain millions of specialized cells for moving nutrients and for reproduction.

Terrestrial algae have adapted to live on land and exhibit robust growth in moist environments such as near waterways or following rain. Algae crusts cover global deserts and get their moisture from morning dew.

What are Microcrop Advantages?

Algae grow on any surface, actively growing or dormant, and flourish on and in the bark of trees and the roots of plants. Legumes use algae in root nodules to harvest nitrogen from the air. Algae and fungi are symbionts in lichen. An algae mat supplies the pigments and food, as well as sugars, while the fungi grow on the outside and protect the algae from desiccation in the sun. Algae also grow in symbiosis with moss, coral, sponges and many aquatic animals.

Terrestrial rooted plants share many similarities with algae because rooted plants evolved from algae 500 million years ago.[238] Early land plants such as mosses had no roots and used algae mats for nutritional support, similar to corals and lichens today.

Roots severely limit terrestrial plants because roots make them immobile and dependent on sufficient moisture and nutrient availability in situ within the rhizome. Land plants die quickly with insufficient moisture because water carries the nutrients needed for cellular metabolism. If too little of any of the 24 essential nutrients are unavailable, land plants die or fail to produce seeds, which defeats reproduction.

Energetic distribution

Algae's photosynthetic mechanism is similar to land plants except algae cells are far more efficient in converting solar energy into biomass. Algae display a simple cellular structure. Cells live in an aqueous environment where they have efficient access to water, CO_2 and can float near the top of the water column to obtain solar energy and nutrients.

The energy available to each cell constrains growth, for both single and multicellular organisms. Available nutrition and solar energy dictate cellular metabolism speed and total energy. An organism can make only so much energy to use immediately for growth or to store for later use.

Rooted crops expend roughly 85% of their energy on non-food growth functions. Therefore, they have only a tiny energy reservoir, and modest energy remaining for food production.

Each energy unit wasted on non-food biomass increases costs for farmers and reduces food production. Food grains lose about 30% of their energy in growing and maintaining roots. Land plants use 35% of their energy supporting a sexual apparatus – tassels, corn kernels, starch and seeds.[239]

Unicellular organisms have the ability to absorb energy and nutrients by diffusion or osmosis in their moist ecosystem. Almost any organic material serves as a potential food source for unicellular organisms. This eliminates the necessity for roots, which consume about 30% of the energy for land plants.[240] Single-celled organisms do not waste energy finding digestible food or expending their energy on a digestive system.

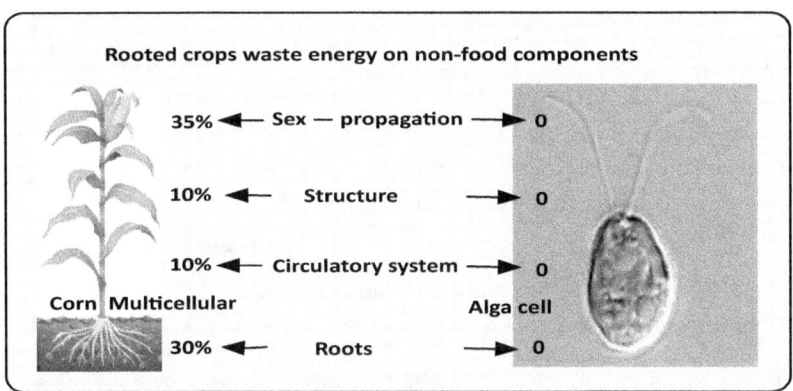

Terrestrial crops such as food grains, above suffer from terrible energetic distribution in a food sense. They must build exotic structures in order to withstand the harsh effects of wind, rain and predators, which saps 10% of their energy. These crops must grow expansive leaves in order to catch photons for photosynthesis and for transpiration. The circulatory and excretory system carries water and nutrients from the roots throughout the plant, which costs another 10% of their cellular energy.

Microcrops are unicellular organisms that have no need for structural components because the cells are supported by water. They have no need for leaves since each cell gathers the solar energy it needs. The cells are so tiny, they absorb nutrients directly from the surrounding water, eliminating need for a circulatory system. Even 90-meter tall kelp lacks a circulatory system as the plant composition is composed of billions of tiny cells that attach together, creating a "pseudo plant."

113

What are Microcrop Advantages?

Many unicellular organisms reproduce asexually, which can take the form of binary fusion, fragmentation or spores. They do not have to waste energy and resources on a sexual apparatus. Many can reproduce both asexually and sexually. Avoiding a sexual apparatus saves 35% of the energy over multicellular plants.[241]

IMA farmers grow staple crops where they have to invest enormous resources to grow food grains that are more than 95% non-digestible cellulose in roots, stalk, stems, leaves, tassels, flowers and husks. Farmers harvest the fruit of the vine, typically the seeds, which may be less than 5% of the crop biomass.

A corn stalk grows slowly supporting two roots; seminal and nodal. It grows 16 to 24 leaves and matures to create a single 640 kernel cob in about 120 days. One ear of corn delivers only 60 calories, 1 g fat and 2 g of protein.[242] Corn produces only during the summer growing season. Slow growth limits food production to a single crop a year.

In contrast, micro crop growers harvest biomass that may be 90% food. Ash, the only non-food residual in algae, accounts for about 9% of the dry biomass.[243] Ash does not go to waste as it is recycled in the culture.

Microcrops enjoy a distinct energetic distribution advantage because they invest nearly all their energy in food biomass. Single-celled organisms waste no energy on superfluous, non-food components. Microcrops offer another and even more powerful competitive advantage over rooted crops, incredibly high-velocity reproduction.

Sex

Algae may be the smartest sexual plants on our planet. When conditions are good, mother cells can propagate by sexual reproduction, involving male and female gametes (sex cells). This allows DNA transfer but slows reproduction. When conditions become difficult, the same algae species may reproduce asexually or use a combination of sexual and asexual reproduction. Asexual reproduction can produce progeny far faster but without the union of cells or nuclear material transfer.

114

What are Microcrop Advantages?

Many tiny algae cells reproduce asexually by ordinary cell division or by fragmentation. Larger algae reproduce by spores. Some red algae produce monospores with walls of agars and carrageenans, long-chained polysaccharides. These spherical cells are carried by water currents and upon germination, produce new organisms.

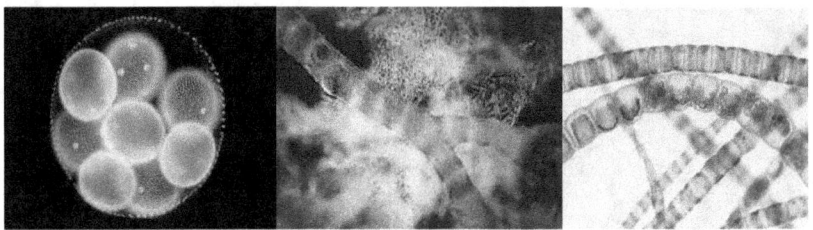

Algae monospores, aplanospores and zoospores

Some green algae produce nonmotile spores called aplanospores, while others produce zoospores, which lack true cell walls.

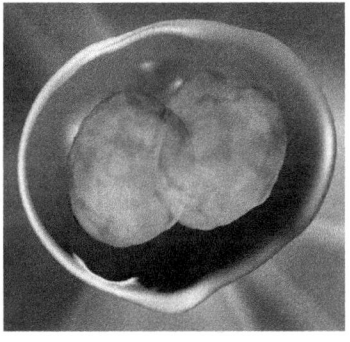

Cell division, mitosis, right, produces a new cell. Asexual reproduction allows single-celled organisms to multiply at very high rates. Many macro and microalgae are asexual and divide by multiple fission. Most share one common feature: under optimal growth conditions, they can divide into two or more daughter cells.

The number of daughter cells produced depends on available light, nutrients and temperature. Cells normally divide twice or three times during a single cell cycle. Cell-cycle progression starts with a period of growth, the G1 phase. During this phase, the cells increase their volume until they roughly double the volume of the daughter cell, coenobia. A coenobia form a colony of alga cells where the cells are arranged in a constant form and number and are surrounded by a gelatinous matrix.

What are Microcrop Advantages?

Algae can create coenobias of up to 16 daughter cells. When the cells have doubled their volume, the cell cycle approaches the commitment

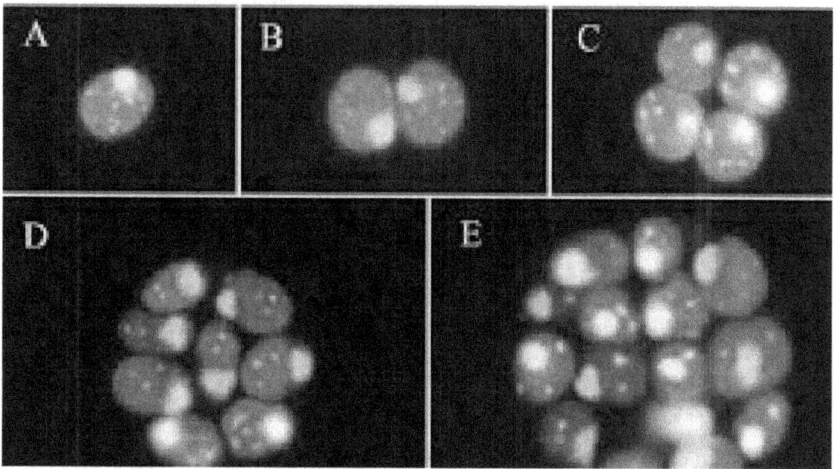

point (CP).[244] When the CP is reached, the cell commits to triggering and then terminating the DNA replication–division sequence, followed by mitosis and cell division.

Scientists are actively studying the CP with the plan to trigger quicker cell cycles.[245] Another accelerated reproduction strategy involves finding a pathway to increase the division number and create more than 16 daughter cells in each cycle.

Mutagenesis gives microcrops another productivity advantage.

Mutagenesis

Algae scientists use algae's ability to adapt quickly to "train" algae with mutagenesis.[246] DNA may be modified, either naturally or artificially, using a variety of physical, chemical and biological agents, resulting in mutations. Scientists examine the mutants to find which best express the desired characteristics. Mutation or variation breeding create crop hybrids among land plants. Successful hybridization may take a decade for land plants because they mature so slowly and then each hybrid must be tested for three generations to insure it passes on the desired traits.

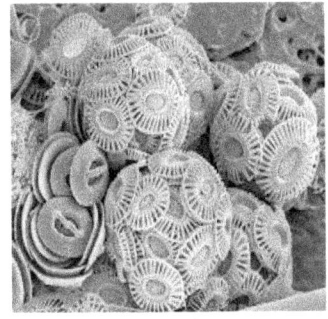

A German team studied the very tiny marine algae Emiliania huxleyii, left.[247] This phytoplankton grows in groups that congeal into large floating masses and serve as food to a wide variety of fish and birds.[248] They chose this algae because it possesses an incredibly fast reproduction rate—500 generations in a single year, nearly two a day.

Algae produce so quickly that mutagenesis often yields a variety of offspring with novel features each day. Phycologists use mutagenesis to modify algae cells for higher production of target compounds, such as specialty oils, long chains of amino acids for proteins or medically desirable ingredients.

Several algae species, such as *Haematococcus pluvialis* react to environmental stress by metabolizing valuable oils, such as astaxanthin. The cells spontaneously produce the oil to protect itself from too much sunlight. Scientists have been able use mutagenesis to triple the percent of astaxanthin found in wild *H. pluvialis*. Astaxanthin can reduce free radicals and oxidative stress and help the human body maintain a healthy state. Astaxanthin provides healthy oils to animal feeds and gives the beautiful pinkish-red color to farmed salmon, shrimp and shellfish.

Salmon, flamingos and shell fish get beautiful color from their algae diet

What are Microcrop Advantages?

Reproduction advantage

Algae reproduce much faster than rooted crops. Algae use photosynthesis to convert carbon dioxide into glucose, which is then broken down into fatty acids. The fatty acids generate membranes for new algae cells, restarting both the reproductive and growth cycles. In some cases, when overloaded with nutrients, algae have such explosive growth, they are called blooms.

With good conditions, some algae species can reproduce in an hour. Freedom food growers plan for several reproductions a day in order to harvest half of the biomass daily.

The single-celled organism reproduction advantage goes far beyond creating valuable biomass 50 times faster than field crops. Developing new field crops by cross-breeding, takes 5 to 10 years or more. New, more productive and valuable algae strains may be developed through mutagenesis or hybridization in a few weeks, since several new generations grow each day.

Algae have been reproducing daily for over 3 billion years, which gives these microcrops significant evolutionary advantage. Most rooted food crops have been hybridized and reproducing once a year, and often less, for only several decades.

Summary

Microcrops enjoy many advantages over their distant progeny, rooted crops. Microcrops do not waste energy on superfluous physical features. Roots are a major difference between fossil and freedom foods. Much of the productivity advantage freedom foods enjoy is the absence of roots.

Microcrops waste no energy and focus only on producing food and bioactive compounds.

Microcrops provide superior nutrition to their many hungry consumers. The most powerful differentiation between fossil and freedom foods, nutralence, is explored in the next chapter.

10. Freedom Food Nutrition Advantage – Nutralence

*Nutrient density defines destiny. Micronutrient deficiencies inflict
>380,000 American infants a year with development disabilities.*

Take aways:

- Why is nutralence critically important for food value assessment?
- How does nutrient density drive destiny?

Freedom foods offer a compelling set of competitive advantages over land plants. Nutralence metrics add to these substantial advantages.[249] Nutralence provides the key to understanding nutritional differences.

Succulents are plants that selectively absorb water. Plants high in nutralence aggressively absorb, manufacture and store proteins and a diverse set of other nutrients. Algae waste no energy on superfluous body functions. Algae focus all their energy on nutrient production. Algae and other microcrops deliver substantially higher nutrient quality, below.

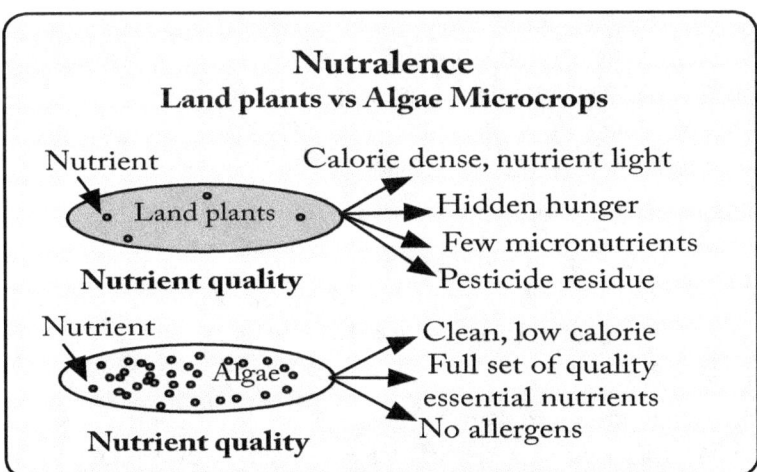

Nutralence includes five critical food compositional attributes: nutrient quality, density, diversity, bioavailability and bioactive compounds.

Functional Foods

People and animals are what they eat. The composition of food plays a significant role in the reachable destiny for children and adults.

Nutrient quality

Nutrient quality covers the full set of macro and micronutrients provided per bite.[250] Land crops have relatively low nutrient quality because the biomass is calorie dense, but nutrient sparse. The lack of micronutrients creates empty calories, nutrient dilution[251] and hidden hunger.[252] Dust, animal dander, house dust mites, pollen, allergens, insects, pesticide and other agri-poisons residues are common and practically unavoidable on field crops and vegetable and fruit produce. These contaminants substantially diminish nutrient quality. Both standard IMA and GMO foods have been bred for yield, which means nutrient quality has been ignored and diminished.

Freedom foods devote their entire energy to nutrient quality. Microcrop foods are clean and do not carry the IMA food residuals. Nature has allowed microcrops to evolve on their own without human interference. As a consequence, microcrops have not compromised their nutrient quality for other attributes like size or weight. Microcrops are cultivated in water and are clean, allergen free and packed with nutrients.

Nutrient density

Each of the five nutralence variables are important for healthy nutrition. **Nutrient density** stands out because microcrops repair the damage done by nutrient deficient field crop staples – corn, soy, wheat and rice.

Freedom foods maximize nutrient density for three reasons.

1. Microcrops grow only to the "limit of nutrients" in the culture. When an alga cell absorbs the last essential nutrient, the plants stop growing. The crop rests or goes dormant for months or years, until the full set of nutrients are available again.

2. Microcrop cells are more than 20 times smaller than field crops. Each cell serves efficiently as a special delivery package providing all the essential nutrients available in each alga cell.

3. Freedom food growers use smart sensors to ensure the culture has the correct nutrient balance at all times. Consistent nutrient balance eliminates hidden hunger and nutrient deficiencies.

Freedom foods provide many benefits. Nutrient density may be #1. Microcrops provide 2 to 20 times more micronutrients per bite with only a few calories. Freedom food nutrient density eliminates empty calories.

Nutrient density defines destiny.

Fossil foods transfer their micronutrient deficiencies to 85% of Americans. Micronutrient deficiencies have resulted in some horribly grim outcomes for families. Children cannot approach their true destiny because, due to no fault of their own, they missed vital nutrients during their fetal life. Nutrient loss during fetal life results in severely under-developed major organs and cannot be undone. In 2018:

- 1 in 10 infants were born preterm, resulting in autism spectrum disorders, developmental disabilities and many infant deaths.
- Developmental disabilities victimized over 380,000 newborns. Painfully for parents, grandparents, family members and America, 1 in 6 newborns will not realize their true destiny due largely to micronutrient deficiencies during womb life.[253]

The Emerald Renaissance plans to stop this horrible loss. The root-cause of nutrient deficiencies in field crops comes from a familiar issue - roots. Terrestrial crops made substantial compromises when they evolved from algae and began their conquest of land.

1. Rooted plants broke through the "limit of nutrients" constraint. They adapted to life on land by developing the ability to grow, mature, and even propagate without a cluster of micronutrients.

This was a useful adaption and allowed plants to survive and prosper. Nutrient availability was far more limited on land than in water. Land plants survived and expanded their territory. The converse of their beneficial adaption has taken on another name today – hidden hunger.

2. Land plants lost the ability to pause or go dormant until sufficient nutrients were restored to the reach of their roots, **and** sufficient soil moisture was available for nutrient uptake.

Terrestrial plants that do not get critical nutrients at the right time stunt or die. IMA farmers address this problem with "precision services" that add to resource consumption, waste and pollution. For example, the crop only absorbs 44% of the substantial fertilizer applied. Macro fertilizers, NPK, do nothing to increase crop micronutrients.

Terrestrial crop nutrient density also suffers from the substantial disadvantage that their nutrient value depends on the micronutrients in the soil that are bioavailable to the roots at each crop life stage. Years of constant IMA crop production strips cropland of micronutrients, organic material and humus, which reduces nutrient quality and density.

Another telling nutrient dilution metric comes from the extremely high-water content of fruits and vegetables. Most fruits contain over 92% water, as do many vegetables. Cucumbers, lettuce, zucchini, tomato radish and celery are comprised of 95% water. Other veggies containing over 90% water include cabbage, cauliflower, broccoli, eggplant, peppers and spinach. After accounting for the considerable non-edible cell wall material, fruits and vegetables simply have little physical material to carry nutrients to their consumers.

IMA crop breeding programs for yield have increased weight, but at the cost of diluting nutrients. Nutrient load has not been a consideration in most plant breeding. Even the food portion of staples such as corn kernels contain 64% non-food biomass, further diluting nutrient density.

The failure of modern staples to carry sufficient nutrient density created the consequence that most consumers do not get the recommended daily intakes of important vitamins and minerals necessary for proper physical and mental development and for child bearing. More than half of American children do not get enough of vitamins D and E, while more than a quarter do not get enough calcium, magnesium or vitamin A.[254] Hidden hunger can result in a compromised immune system, stunted physical growth, reduced mental ability, chronic disease and premature disability and death.

Functional Foods

Three nutrient dilution studies on fruits and vegetables examined by Donald Davis at the University of Texas Biochemical Institute, examined historical food composition data. He found declines of 5% to 40% or more.[255] Another study evaluated vitamins and protein dilution over the last 20 years with similar results.

Nutrient dilution in IMA foods has increased with cropland damage from soil compression, erosion and fertility loss. Rooted crops cannot store micronutrients in their fruit unless those micronutrients are available in the soil, water soluble and within reach of the roots. Systemic fertility loss adversely impacts nutralence, especially nutrient density.

Microcrops deliver 10 to 100 times more nutrient density than rooted crops, (below.) Field crops distribute 85% of their energy to non-food elements like roots, which diminishes energy to produce nutrients.

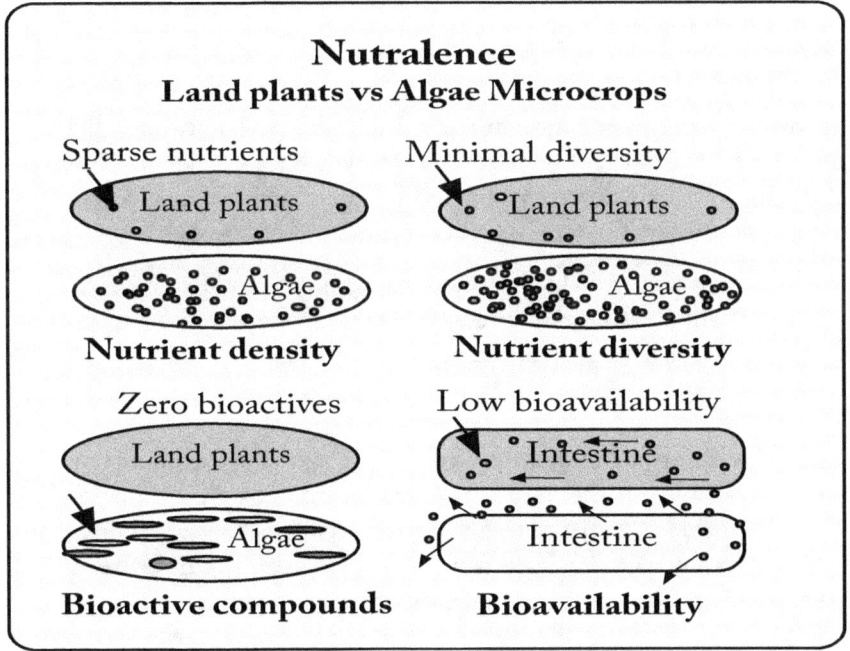

Tiny cell size plays an outsized role in nutrient density. Corn and other multicellular organisms have relatively large cells where nutrition is sparse and distributed widely. In contrast, microcrops have all their nutrition packed into a very tiny cell, 5µ to 10µ (microns). Nano cell size results in massive surface area for nutrient absorption. A cubic container of algae cells the size of a postage stamp, 2.2 cm by 2.5 cm, has the surface area of three football fields. The tiny container holds billions of algae cells but only several thousand of field crop cells like larger corn fit in the container.

The nutralence in Spirulina is evident when compared with terrestrial foods with the highest levels of key nutrients.[256] Bite for bite, Spirulina has 180% more calcium than milk, 670% more protein than soy tofu, 3,100% more β-carotene than carrots, 2,200% more vitamin C than tomatoes, and 5,100% more iron than spinach.[257] Microcrop foods can provide 77 different macro and micronutrients, vitamins, trace elements and over 200 bioactive compounds, most of which are not found in field crops. Rooted crops deliver lots of weight but generally provide only a small fraction of nutrient density delivered by microcrops.

Nutrient diversity

Nutrient diversity plays an important role in health and nutrition. IMA monocrops have little nutrient diversity since only 12 crops make up 75% of the world's food.[258] Crop breeders for food staples such as corn and soy sacrificed nutrient diversity for yield with every new crop developed.

While yield improvements helped farmers with bigger produce, it severely diminished nutrient density and diversity. Microcrops offer thousands of species alternatives, each with splendid nutrient diversity.

Nutrient bioavailability

Nutrient bioavailability represents the fraction of ingested food nutritional components available at the target for use in physiological functions. Bioavailability includes both nutrient bioactivity and bioaccessibility.[259] Bioaccessibility refers to the release from the food matrix, transformations during digestion, and transport across the digestive epithelium.

Bioavailability entails the entire process including: digestibility and solubility of the food in the gastrointestinal tract; absorption or assimilation of the food element across the intestinal epithelial cells and into the circulatory system; and finally, incorporation into the target site of utilization.[260]

Eating good food is a waste of energy if the nutrients are not bioavailable and absorbed in body tissues. Farmers are quite aware that meat and dairy animals find corn hard to digest because their stomachs did not evolve or adapt to eating field grains. Antibiotics aid digestion, which is why 80% of the antibiotics in the US are used in animals.[261]

Unlike corn and soy that are relatively hard to digest, algae-based foods are easy to digest, which improves bioavailability. Again, size matters. Algae cells are so tiny and move easily through the intestine wall to the bloodstream. Rooted plant cells are 20 times larger and many of the nutrients may simply pass through the body unabsorbed. Some algae cultivars require pre-processing to break down strong cell walls in order to achieve high bioavailability.

Animal studies with algae biofeeds for fish and cattle show these animals create less liquid and solid waste compared with field grains. Algae biofeed delivers nutralence in such a minute package that the nutrients are immediately bioabsorbed, creating less pass-through waste.[262]

Bioactive compounds

Bioactive compounds may be the most important nutritional difference separating microcrops and rooted crops. These diverse compounds play a vital role in sustaining health and avoiding disease naturally.

Terrestrial crops are similar to Ole' Blu, a Tick Hound that came from a fine linage of great hunting dogs. Blu was raised in an Atlanta condo and never saw a field. About age 6, his owner took him hunting. Ole' Blu did not have a clue about what to do. He had lost his keen sense for hunting. Field crops have been carefully tended by farmers for thousands of years. They have lost the need for bioactive compounds.[263]

Bioactive compounds are produced in plants as secondary metabolites.[264] Both macro and microalgae, are rich in bioactive antioxidants, soluble dietary fibers, proteins, minerals, vitamins, phytochemicals, and polyunsaturated fatty acids. Many algae species have therapeutic properties that improve health and prevent or treat disease.[265]

Izabela Michalak and Katarzyna Chojnacka published an excellent review article describing the diversity of bioactive compounds algae produces.[266] Bioactive compounds provide both protection from health threats, (viruses and bacteria), as well as therapeutic treatments.[267]

Primary bioactive compounds in algae	Bioctive compounds for degenerative metabolic disorders:
• Antioxidants • Soluble dietary fibers • Proteins and amino acids • Minerals and trace elements • Vitamins and phytochemicals • Polyunsaturated fatty acids	• Sulphated polysaccharides • Phlorotannins • Vitamins and minerals • Carotenoids (e.g. fucoxanthin) • Peptides and sulfolipids
Improve health, prevent or treat disease	**Other bioactive compounds:**
• Anticancer • Antioxidant • Antiobesity • Antidiabetic • Antihypertensive • Antihyperlipidemic, • Anticoagulant • Anti-inflammatory	• Immunomodulatory • Antiestrogenic • Thyroid stimulating • Neuroprotective • Antiviral • Antifungal • Antibacterial • Tissue healing

Bioactive Compounds

Sea vegetables, macroalgae — Green algae — Blue-green algae, cyanobacteria

Bioactive compounds are not essential for the daily functioning of the plant, such as growth. They play many other significant beneficial roles in competition, defense, repair, attraction and signaling. The plant's bioactive compounds pass the bioactive benefits to their consumers, which may be another plant, animal or human.

Functional Foods

Bioactive algae compounds include proteins, sulphated polysaccharides, phlorotannins, carotenoids, minerals, peptides and sulfolipids, with proven benefits against degenerative metabolic disorders.[268]

They are called bioactive because they use ingenious biological strategies to protect cells and to block invasive pathogens. Algae evolved over eons in extremely harsh environments. The robust cells that emerged depended on the presence of bioactive compounds for their survival. Bioactive compounds stimulate cellular metabolism in ways that cause the body's immune system to produce enzymes, hormones, peptides or other compounds that disrupt the sequence of a malady's onset. Different algae types – sea vegetables, green and blue-green algae – provide bioactive compounds in different proportions.

What are actual nutrient levels?

Direct comparison of freedom foods and the best fossil crops provides a good way to understand microcrop nutralence levels.

Microcrop nutrient levels vs popular foods per kg	
Nutrient	Microcrops have _ times higher than
Proteins - building blocks for bones, muscles, cartilage, skin, and blood. They help manufacture enzymes, hormones, and vitamins.	2x > than soy 3x > than beef, fish, pork 6x > than eggs
Iron carries oxygen in the blood. Many women in childbearing years have iron-deficiency anemia.	30x > than beef, fish, pork 65x > than spinach
B12 vitamin helps the body release energy, regulates the nervous system, aids in the formation of red blood cells, and help build tissues.	3 to 4x > than animal liver
Magnesium heart rhythm, build bones, maintains the immune system and normalizes blood pressure.	2x > than spinach 5x > than tomatoes
Calcium regulates nerve transmission, blood clotting, hormone secretion and muscle contraction.	10x > than milk
β-carotene, pro-vitamin A, boosts immune system, helps skin, eyes, protects against CHD and cancer.	5x > than carrots 40x > than spinach
Chlorophyll helps fight cancer, speeds wound healing, cleanses liver of toxins, improves skin, digestion and weight control.	30x > than spinach 20x > than wheatgrass
EPA and DHA, omega-3 fatty acids improve eyesight, brain function and protect from CHD.	1,000x > than any land plant 1,000x > than beef, poultry

Functional Foods

Protein acts as a vital building block for body functions. One kilo of a microcrop like algae delivers twice the protein of soy, 2.5 times more protein than animal meat and six times more than eggs.

Algae produce pigments, such as carotenoids, (carotene, xanthophyll), chlorophylls and phycobilins, (phycocyanin, phycoerythrin). These pigments not only have nutritional value, but also contain a variety of antibacterial, antiviral, antifungal, antioxidative, anti-inflammatory, and antitumor properties. Algae manufacture large quantities of antioxidants, polyphenols, tocopherols, vitamins and mycosporine-like amino acids that provide vital cellular protection and repair.

Research on bioactive compounds remains in its infancy, similar to understanding gut microbiota.[269] Microflora in the gut use bioactive compounds to manufacture enzymes, hormones, peptides and other elements that improve health and fight pathogens. The benefits from eating freedom foods will multiply with advances in bioactives R&D.

Emerald Protein

All protein is not created equal. Fossil plant and animal meat protein suffers from a lack of nutralence.

Emerald Protein Delivers Higher Nutralence

Nutralence	Emerald protein compared to fossil plant or meat protein	Criticality
Nutrient quality	10x to 100x	No contaminates. No pesticide residuals.
Nutrient density	10x to 500x	No empty calories. More nutrients per bite.
Nutrient diversity	100x to 1000x	More micronutrients, vitamins and trace elements.
Nutrient bioavailability	10x to 500x	Faster and more reliable assimilation into the body.
Bioactive compounds	100x to 1000x	More disease protection. More disease therapeutics.

Functional Foods

Fossil foods have been bred to maximize protein. Little thought was given to 80 other macro and micronutrients, vitamins minerals and trace element that support cellular metabolism. Fossil foods also miss over 200 valuable bioactive compounds.

Freedom foods produce **emerald protein** which delivers the full set of macronutrients micronutrients in a tiny, bioavailable nano-cell package. On an emerald protein basis alone, many consumers will prefer freedom foods over field crops or animal meat. Consumer choice will become even stronger when algae-based meats are available with the taste and texture consumers crave in beef, pork, lamb, goat, poultry and seafood.

The nutralence difference – microcrops versus field (rooted) crops – show significant differences. These metrics, e.g. 10 times superior nutrient quality, need to be independently validated over the next decade by food scientists. No terrestrial foods currently offer the key nutralence metrics today, probably because those numbers would be incredibly poor. Fossil foods' failing nutralence metrics should not be ignored because they are the reason for the fossil food-caused the epidemic of preterm births, obesity, diabetes and Wester diseases.

For consumers who understand that "nutrient density drives destiny," and are interested in maximizing health with the most nutrients per bite, microcrops offer a substantially better nutrition option than fossil foods.

Why the fossil food protein disadvantage?

Terrestrial plants made two enormous sacrifices after they evolved from algae. One came from nature and one caused intentionally by humans. Land required plants to support several new energetically demanding features – roots, stems, leaves, circulatory system and a sexual apparatus.

A plant has only the limited energy it can capture from the sun. How the precious solar energy becomes distributed determines its growth and food production efficiency. Land plants sacrifice efficiency in capturing photons from the sun because existing cells are consistently shaded by new growth. The percent of cells on a corn plant shaded from solar energy is probably >10,000 times more than the cells facing the sun.

129

Functional Foods

Rooted crops ability to grow protein efficiently fell sharply with the addition of roots, because the plant had to distribute its limited energy among competing tasks, most of which had nothing to do with food production. Roots demanded energy for to go deeper for moisture and still more energy to pump nutrients from the roots through the plant. Stems commandeered energy as they made demands for vertical growth and strength to survive severe environmental stresses and storms.

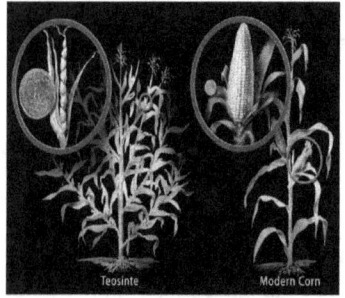

Corn provides a good case study. The original teosinte plant, left, the parent of modern corn, probably had higher nutralence than corn today. Early farmers and later the Incas and Aztecs selectively bred teosinte to improve plant and kernel size but ignored nutrition because nutritional metrics were not measurable.

The sacrifice created by human-directed hybrids was simple but also had a significant impact on protein production and total nutralence. Farmers' primary goal has been higher yields in order to produce more food. Farmers are smart and saved seeds from the strongest, largest and most colorful plants for next year's crop.

Plants were bred for size or yield weight, with almost no consideration for nutralence. For nearly all of human history, food nutralence was neither visible nor measurable. The next generation of consumers will change the status quo and demand high-nutralence.

Tipping point

Coming soon are a suite of nutrient scanners that consumers can use to help make food decisions by identifying nutrients and nutrient density in their food. Multiple technology companies are developing handheld scanners using spectroscopy, which astronomers have used for decades to detect elements in distant stars.

Functional Foods

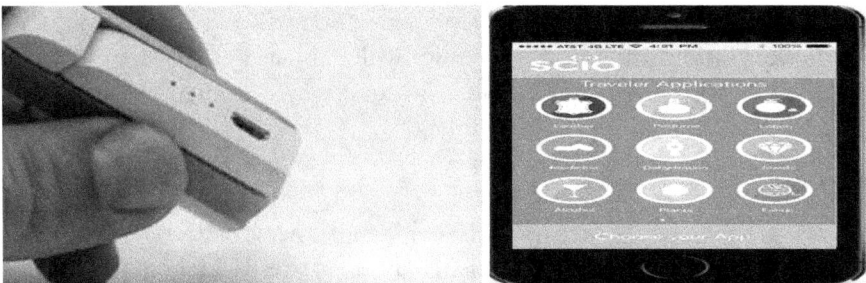

The SCiO scanner is under development by Consumer Physics based in Tel Aviv, Israel. SCiO detects the molecular food signature and sends the data to a smartphone. SCiO's database uses algorithms that translates the molecular signature into nutritional content. Consumer behavior changes rapidly when the **invisible become visible**.

China smog proves the point. Recently, handheld scanners became available in China for detecting the quality and density of air pollution. People immediately changed their habits because they could avoid the most dangerous and toxic exposure by timing their time outside to minimize black soot particulates, PM2.5. People encouraged their government to change energy production from coal to natural gas. A 2018 study that incorporated the use of smog scanners found that people were able to reduce their PM2.5 exposure by 47%.[270]

Imagine when people can choose healthier foods with clear metrics for nutrient quality, density, diversity, bioavailability and bioactive compounds. Some people will embrace the new foods to avoid micronutrient deficiencies and contaminants. Others will adopt freedom foods for broad or specific health benefits.

Every semester my food marketing students would push the urban legend that large food companies intentionally market puff, big foods with nutrient deficiencies. Rather than tell them that my consulting experience with most the major food companies indicates the opposite, (which they would not believe), students were given an assignment. They were tasked to use the internet to find and analyze in a one-page memo the vision and values for five companies in the food supply chain.

Functional Foods

Those memo results over 30 years are very compelling. Every student came to the same conclusion: "Big food companies will embrace and market high nutralence foods when they are available." Scanners will change the food landscape for food companies and consumers.

Summary

A new set of nutralence metrics will help consumers understand the nutritional value in their food.[271] Nutralence differentiates microcrops from fossil foods – both plants and meat. These metrics will test the degree of difference provided by any novel food alternative. The nutralence model provides a path for food technology scientists to create metrics that benefit all consumers.

Nutralence metrics expose fossil foods nutritional weaknesses, especially compared with freedom foods. While nutrient density may be most important today, access to bioactive compounds may have the most value in the future because they deliver so many benefits.

Nutralence dimension	Food score: Fossil	Freedom
• Nutrient quality – low quality, contaminants	1	10
• Nutrient density – very low, hidden hunger	1	10
• Nutrient diversity – extremely low, almost none	1	10
• Nutrient bioavailability – low, hard to digest	1	8
• Beneficial bioactive compounds – none	0	10

A food supply system should positively support consumer health. IMA fossil food production predisposes consumers to the full array of Western diseases, which argues for a new, healthier food system.

Grocery stores are likely to conveniently place hand scanners that report a consumer's personal micronutrient levels. The consumer may pass her hand over the scanner and receive a private report on her smart phone.

The March of Dimes may promote: *Nutrient density defines destiny,* and provide hand scanners, especially to mothers of child bearing age. The March of Dimes Prematurity Campaign aims to reduce premature birth in the US and to give every baby a fair chance for a healthy full-term birth.[272]

11. Functional Foods

Functional foods are the engine that delivers desired target compounds that benefit vital body functions.

Key takeaways

- What are the functions functional foods provide?
- How does microcrop nutralence support functional foods?
- How do bioactive compounds life functional foods?

Ancient medicine – Chinese, Egyptian, Mayan and Greek – focused on restoring the fine balance of energy, body, and spirit to maintain health. Medicines were developed over centuries by trial and error. Therapeutic extracts were made from parts of plants – often leaves, bark or roots – with the goal of restoring health by nourishing the body.

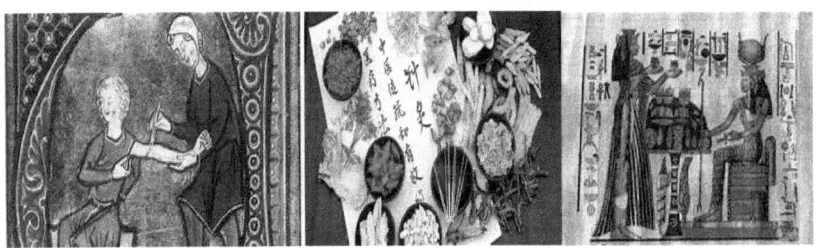

Ancient cultures wisely embraced bioactive compounds to sustain health.

Functional foods share similar features as they deliver specialty compounds that beneficially affect one or more body functions. Functionals enhance the immune system to fight off infections and a broad spectrum of diseases. Several offer anti-inflammatory properties that moderate diseases like CHD, arthritis, asthma and Alzheimer's. These foods use herbs or algae extracts to provide bioactive compounds such as omega-3, that transfer their beneficial functions to consumers.

Functional Foods

Microcrops prevent disease

Microcrops offer novel biosolutions. Microcrops such as algae are photoautotrophic cell factories that provide an efficient means of converting solar energy into biomass composed of fatty acids, lipids, vitamins, carbohydrates, antibiotics, antioxidants, proteins and bioactive compounds that can battle disease.

Microcrop compounds prevent disease by providing the essential nutrients for sustained health and vitality.[273] These plants contain high multiples of the nutrients that cause the major nutrient deficiencies.

Algae contains several times more beta-carotene, (provitamin A) than other foods. Algae are rich in antioxidant vitamins, (C and E), in concentrations far higher than any land plants. Vitamin C provides protection against immune system deficiencies, CHD, prenatal health problems, eye disease, and skin wrinkling. Vitamin E moderates neurological problems due to poor nerve conduction and anemia due to oxidative damage to red blood cells. Algae are a good source of all seven B vitamins. Algae offer a unique as a plant source of vitamin B12.

Algae provide a mineral profile superior to that of land plants, milk, eggs or soybeans without allergens. Terrestrial foods such as food grains have minerals such as iron bound up in phytic acid complexes, limiting their bioavailability.[274] These complexes cannot be absorbed into the blood stream and simply pass through the body without being absorbed. Studies show iron absorption significantly higher for algae compared to rice, other food grains or beef.[275]

Algae are rich in iodine and selenium, critical trace elements that are highly variable in food supplies by geographic region.[276] These minerals have been associated with endemic deficiency disorders throughout history. Algae concentrate these trace minerals in one tablespoon of dried algae, which provides sufficient levels of these nutrients when introduced into the diet. Algae have a high content of glutamic acid that stimulates taste receptors with umami, (savory or hearty).[277] Umami amplifies taste differentiation and increases the desire to consume algae for taste.

Functional Foods

Bioactive compounds

Bioactive compounds operate individually, or more often symbiotically, to assist the body's defenses and to protect from threat vectors such as viruses and bacteria, (antiviral and antibacterial). When the body does become infected, they work to neutralize or expel the pathogens, (e.g. anticancer) and repair tissues and organs.

Microcrop foods provide a diverse array of over 200 bioactive compounds. Bioactives are prevalent in both macro and microalgae. Izabela Michalak created a table of useful macroalgae compounds that serve multiple functions. Izabela and her team have published a series of excellent papers describing sources, uses and therapeutic benefits of many bioactive compounds, below.[278]

Function	Bioactive algae compounds
Antibacterial	Proteins, polyphenols, PUFAs* polysaccharides, pigments: chlorophyll, carotenoids.
Antifungal	PUFAs, pigments: chlorophyll, carotenoids, terpens, phenols.
Antioxidative	Proteins, mycospoine-like amino acids, PUFAs, carotenoids, tocopherol, ascobaate.
Anti-inflammatory	Proteins, PUFAs, carotenoids, polysaccharides, sterols – fucosterol, polyphenols – phlorotannins, porphyrin derivati, hoephorbide and pheophytin.
Anti-tumor	Polyphenols, carotenoids, polysaccharides.
Antiviral	Proteins, diterpens, polyphenols, polysaccharides, carotenoids.

Properties of algae bioactive compounds – Izabela Michalak
*Polyunsaturated fatty acids, PUFAs such as omega-3.

135

Functional Foods

Common additions to functional foods today are long chain omega-3 fatty acids-DHA/EPA, flavones, beta-carotene, lutein lycopene, fiber, catechins, and anthocyanins. Smart food companies are adding precisely the nutrients that have caused the most serious nutrient deficiencies – Vitamin A, B complex, D and E along with magnesium, zinc and iron. Flavonoids, one of 200+ bioactive compounds, are phytonutrients that contribute to color.

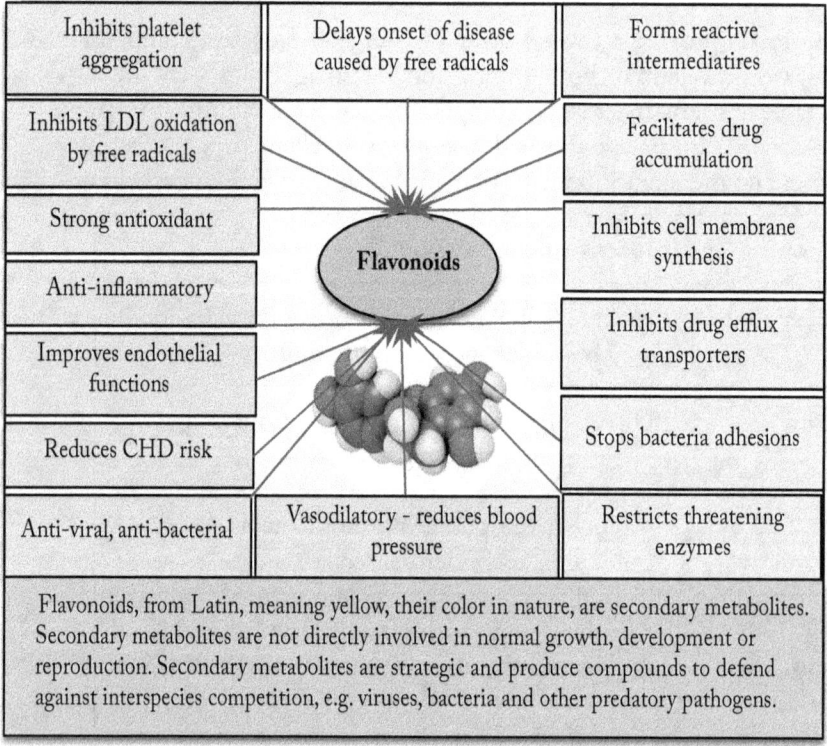

Inhibits platelet aggregation	Delays onset of disease caused by free radicals	Forms reactive intermediatires
Inhibits LDL oxidation by free radicals		Facilitates drug accumulation
Strong antioxidant	**Flavonoids**	Inhibits cell membrane synthesis
Anti-inflammatory		
Improves endothelial functions		Inhibits drug efflux transporters
Reduces CHD risk		Stops bacteria adhesions
Anti-viral, anti-bacterial	Vasodilatory - reduces blood pressure	Restricts threatening enzymes

Flavonoids, from Latin, meaning yellow, their color in nature, are secondary metabolites. Secondary metabolites are not directly involved in normal growth, development or reproduction. Secondary metabolites are strategic and produce compounds to defend against interspecies competition, e.g. viruses, bacteria and other predatory pathogens.

Flavonoids provide antioxidants, which play a significant beneficial role in cardiovascular health and may help to prevent against diseases caused by free-radical damage.

Food buyers choose functional foods to improve their health and well-being and to reduce disease risk. Freedom foods support functional foods with the nutralence and distinct bioactive compounds that improve health and vitality while protecting against disease.

Functional Foods

Decades of consumer research shows people want to eat highly nutritious and enriched foods close as possible to the food's natural state. Functional foods take many forms, including cereals, breads, bars, soups, stews or beverages that are fortified with specific nutrients, vitamins, herbs, and nutraceuticals.

Functional foods have been available only at health food stores for decades. In the last few years, they have exploded across the food isles of regular grocery stores. This new food category has become so important, two new scientific journals were recently born covering only functional ingredients.

The *Journal of Functional Foods* serves as the official scientific journal of the International Society for Nutraceuticals and Functional Foods. *Functional Foods in Health and Disease* is a peer-reviewed journal.[279]

A library search for "algae and functional foods" returned over 900 articles. Market studies estimate the global sales for functional foods will eclipse $255 billion by 2024.[280] When consumers realize the value of functional foods, that estimate is likely to double.

The XTC World of innovation, the most qualitative worldwide database of FMCG, (Fast Moving Consumer Goods), published a global database indexing innovative food products on the market. XTC reported several new food products containing algae launched in 2015–2016.[281] These include sea vegetable crisps, enriched milk-based powder, sea vegetable biscuits, algae instant mashed potatoes, tagliatelle and Wakame salad.

The first comprehensive text on functional ingredients appeared in 2013.

Herminia Dominguez, edited the comprehensive Functional Ingredients from *Algae for Foods and Nutraceuticals*.[282] The International Food Navigator launched an algae special edition newsletter in August 2016, highlighting new algae products and their derivatives.[283]

Functional Foods

Why the excitement?

The *Functional Foods: Key Trends & Developments in Ingredients* 2018 report predicted "Microalgae to be the most exciting functional food ingredients, showing great promise."[284] The report projected algae protein, omega-3 fatty acids, vitamin D, magnesium and whole food microalgae would be the key algae ingredients incorporated in functional foods. These ingredients have propelled the health-food market for years. Now they are entering the mainstream consumer food market.

Top food marketers target Millennials. These consumers, ages 14 to 33, view their food choices as healthier, more expensive, more natural/organic, less processed, better tasting and fresh.[285] Millennials are also the most likely to believe that functional foods and beverages can be used in place of some medicines to relieve tiredness and lack of energy, retain mental sharpness with aging, and manage stress and eye health.[286]

The *Journal of International Food Technology* reported that 80% of consumers believe that functional foods can help prevent or delay the onset of heart disease, hypertension, osteoporosis and Type 2 diabetes.[287] Six in ten consumers associate functional foods with benefits linked to age-related memory loss, cancer and Alzheimer's disease. In 2017, 56% of consumers bought foods or beverages that targeted a specific condition, especially fat and cholesterol-lowering foods and drinks.[288]

Extra-nutritional constituents

Bioactive compounds in modern foods are extra-nutritional constituents that occur in small quantities in foods. Medical scientists are studying bioactives intensively to assess multiple health effects. Numerous epidemiologic studies have shown protective effects of plant-based diets high in bioactive compounds on obesity, diabetes, cardiovascular disease (CVD) and cancer.[289] These compounds vary widely in chemical structure and function, for example:

- **Phenolic** compounds and flavonoids are present in plants and have been studied in cereals, legumes, nuts, olive oil, vegetables, fruits, tea, and red wine. Phenolic compounds have antioxidant properties, and some studies have demonstrated favorable effects on thrombosis and tumorogenesis. Some epidemiologic studies have reported protective associations between flavonoids or other phenolics and, osteoarthritis, CVD and cancer.[290]

- **Phytoestrogen** compounds have antioxidant properties and are present in olive oil. Some studies demonstrated favorable effects on other CVD risk factors, and in animal models of cancer.

- **Resveratrol,** found in nuts and red wine, has antithrombotic and anti-inflammatory properties and inhibits carcinogenesis.

- **Lycopene,** a potent antioxidant carotenoid in tomatoes and other fruits may protect against prostate and other cancers and inhibits tumor cell growth in animals.

- **Organosulfur** compounds in garlic and onions and isothiocyanates in cruciferous vegetables have anticarcinogenic actions and cardioprotective effects.

Numerous bioactive compounds have beneficial health effects. More scientific research needs to be conducted before producers can make science-based dietary recommendations. There is sufficient scientific evidence to recommend consuming food sources rich in bioactive compounds.

Today, the best way to consume some bioactive compounds is the Mediterranean diet rich in a diversity of fruits, vegetables, whole grains, legumes, olive oil, and nuts. Soon, freedom foods will provide bioactive compounds that will pass their beneficial health effects to consumers.

What are the functions?

Algae act as a natural biofactory of valuable extractable compounds to serve as ingredients in functional foods. Algae biomass contains carbohydrates, proteins, minerals, oil, polyunsaturated fatty acids and over 200 bioactive compounds. A target bioactive compound may be found as only 0.2% of the biomass in a wild species. Phycologists use mutagenesis and other microtechnology tools to "train" the cells to over-produce the desired compound to possibly 2% of the biomass.

Functional Foods

Hundreds of scientific articles describe the functions from algae-based bioactive nutrients and explain the mechanism for action. For example, algae antivirals use several ingenious strategies to attack or defeat viruses.[291] Some bioactive compounds make cell walls like Teflon, so the viruses cannot stick and do their damage. Others guard the DNA and RNA, protecting the cell signaling from infection.[292] Some block viruses from releasing its genetic material or disrupt their functions.

Another strategy inhibits the virus from reproduction. Algae can biostimulate the cell to produce macrophages that engulf the virus.[293] The body senses the surrounded virus as a waste material and sluffs it off naturally from the body. A final strategy targets the virus indirectly, by increasing the efficiency with which the host's immune system can fight the viral infection.[294]

Why algae?

The Natural Products Expo, East and West, each attract 100,000+ visitors that examine hundreds of booths displaying the benefits of natural foods and extracts. Dozens of purveyors offer bioactive compounds extracted from leaves, barks, seeds and roots sourced from exotic locations around the world. Without a nutrient scanner, buyers have no idea about the quality, density or bioavailability of each offering.

Algae biomass produced in a controlled environment will become the primary source for bioactives for five reasons; density, quality, consistency, bioavailability and cost. Algae provide substantial advantages over rooted plant sources because algae offer:

- A substantially higher percentage of the target bioactive in the plant biomass, making extraction less tedious and expensive.
- Higher quality, cleaner compounds free of contaminates. Harvests in natural settings often includes many undesired compounds.
- Higher target compound consistency. Rooted crops display wide variation based on harvest timing, geography and microclimate variation. Microcrops provide consistent target compound quality.
- Significantly higher bioavailability due to the small cell size.

- Lower cost for growers and producers, which should transfer to consumers.

Regulatory and product quality issues for functional foods and food ingredients have been reviewed by the FDA, USDA, multiple other countries and the European Union.[295] Regulations vary substantially among different countries and cultures. Most the regulations focus on the type of food or nutrients, not nutrient quality, density, diversity, bioavailability or bioactive compounds.

Taste

Consumers will leave functional foods on the shelf unless they taste good. Food consumers are clear – taste rules! Their primary concerns about algae foods are taste, texture, color, aroma and mouth feel.

Considerable market research shows consumers are unwilling to sacrifice taste for health benefits. Wim Verbeke, from Belgium, did a series of studies that concluded the health benefit beliefs from functional foods emerged as the strongest positive determinant of willingness to compromise on taste.[296] Both the level of perceived benefits and its predictive power on willingness to compromise on taste decreased over time. Related research has shown that when people create high expectations for health or other attribute, their satisfaction diminishes quickly if they do not see, feel or taste benefits.

Microcrops provide of all the essential nutrients, vitamins, minerals, and trace elements essential for health and vitality with only a few ounces a day.[297] Algae can satisfy any appetite with a broad spectrum of aromas, colors, tastes, and textures. Algae have a high content of glutamic acid that stimulate taste receptors, amplify taste differentiation and the desire to consume algae for its good taste. Most functional food studies try to develop new practical properties of algae nutrient extracts that enhance food palatability. A beef patty texture is improved by the addition of 3% *Wakame* powder.[298]

Addition of algae in pasta reduces cooking loss without alteration of the sensory attributes.[299] In the beer industry, sugar kelp found a place in beer, enhancing the malty taste.[300] Other algae components improve liquid clarity. Laurie-Eve Rioux with the Canadian Institute on Nutrition and Functional Foods as done excellent work on the value of functional foods. She provides a comprehensive table with the algae species used, and a link to the producing company's website.

Some sea vegetables are rich in sugar alcohol, which contains about 30% mannitol. When the plant fronds are soaked in hot water, mannitol releases in the broth.[301] Mannitol provides a sweet taste without the calories from sugar. Kombu, mannitol and glutamate, (in the form of monosodium glutamate) builds layers of the umami and sweet flavor in broth. The combination of mannitol with glutamate may open up to different flavoring profiles. Taste tests made by multiple chefs indicate that the palatability of low-fat food is improved by umami flavors.[302] Umami taste improves not only the food with the umami flavor, but also enhances flavors for other foods in the meal.

The next chapter examines how freedom foods can prevent and restore health from Western diseases.

12. Health Restoration from Fossil Diet Diseases

Unhealthy diets now pose a greater risk to morbidity and mortality than unsafe sex, alcohol, drug and tobacco use combined.[303]
                                    - EAT Lancet: Food, Planet, Health, 2019

Key takeaways

- Can freedom food solutions prevent or treat Western diseases?
- How do microcrop solutions work for obesity and diabetes?
- What Western diseases can algae-based medicines address?

Americans made a Faustian deal with the devil – asking for cheap food – which IMA delivered. Initially unseen in the exchange, hidden hunger, came from stealthy micronutrient deficiencies. This led to hundreds of deadly diseases, agonizing disabilities and widespread premature death.

The cheap fossil foods diet forces Americans pay a steep price, poor health. Americans spend twice as much per person on healthcare than people in other rich nations, yet die younger and endure more injury and illness.[304] In 2018, 72% of American adults are overweight or obese.

The National Institute of Health discovered US men ranked last in life expectancy among 17 wealthy countries and American women ranked next to last. The US also ranked at or near the bottom in nine chronic health areas, including heart disease, lung disease, obesity and diabetes, injuries and homicides and sexually transmitted diseases.

Substantial biomedical and clinical evidence suggests that the primary cause of the current Western disease epidemic is the Western diet of fossil foods.[305] Consumption of modern processed foods with high fat, cholesterol, protein, sugar, and salt, as well as dairy and animal meat products, promote a myriad of chronic diseases, below, throughout each life stage.[306]

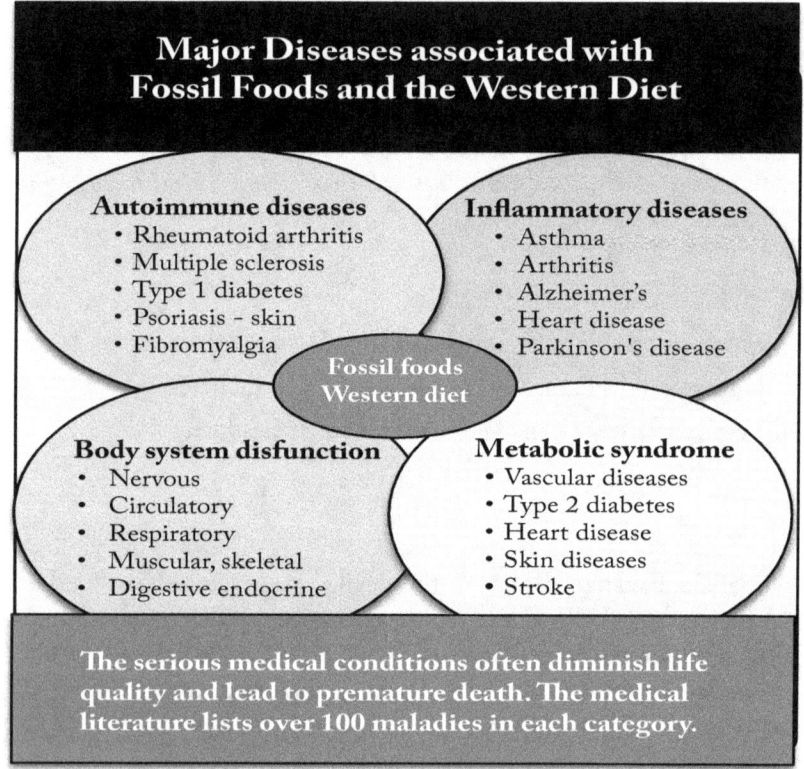

Major Diseases associated with Fossil Foods and the Western Diet

Autoimmune diseases
- Rheumatoid arthritis
- Multiple sclerosis
- Type 1 diabetes
- Psoriasis - skin
- Fibromyalgia

Inflammatory diseases
- Asthma
- Arthritis
- Alzheimer's
- Heart disease
- Parkinson's disease

Fossil foods Western diet

Body system disfunction
- Nervous
- Circulatory
- Respiratory
- Muscular, skeletal
- Digestive endocrine

Metabolic syndrome
- Vascular diseases
- Type 2 diabetes
- Heart disease
- Skin diseases
- Stroke

The serious medical conditions often diminish life quality and lead to premature death. The medical literature lists over 100 maladies in each category.

Freedom food solutions

Microcrop bioactive compounds formulated in functional foods, nutraceuticals, pharmaceuticals and medicines can address each of these disease areas. Microcrop medicines provide therapeutic solutions that can assist millions restore good health.

Medical scientists are conducting lab, animal and human trials for restorative algae-based medicines in each of the major disease health areas. A PubMed or ProQuest search using "algae" and "insert a chronic condition," typically returns dozens of articles. A search for "algae" and "cancer" articles returned 564 hits, diabetes 210, inflammation, 228, and immune function, 244. Extensive medical research with algae bioactive compounds can be found in my blogs, Algae101 and AlgaeSecrets.[307]

Health Restoration from Fossil Diet Diseases

Freedom foods use two strategies to prevent or treat serious health conditions.

1. **Restore micronutrient health** uses the whole high-nutralence algae food to deliver the full set of micronutrients and bioactive compounds that repairs nutrient deficiencies and restores health.

2. **Leverage specific bioactive compounds** extracted from microcrops and used for targeted therapeutic solutions such as disrupting the events cascade for asthma or halting the spread of cancel cells.

Freedom Foods Diet **Whole, High Nutralence Natural Foods**		
Whole foods	**Food types**	**Strong nutralence**
• Low fat	• Whole foods	• High fiber
• Low cholesterol	• Plant-based dairy	• High nutrient quality
• High protein	• Plant-based meat	• High nutrient density
• Low sugar	• No GMO	• High bioavailability
• Low salt	• No pesticide residuals	• High bioactive compounds

Restoring micronutrient health provides the easiest strategy because consumers can overcome nutrient deficiencies by eating any of many food choices. The examples here, women of reproductive age and their children, illustrate the value proposition that works equally well for people of all ages.

Restore micronutrient health

Freedom foods alone without additional bioactive compounds can resolve an array of major health issues for the unborn, infants and young children. The high-nutralence foods deliver precisely the essential micronutrients to attain and sustain good health.

Women of reproductive age and their children experience devastating health consequences from nutritional deficiencies. Women pass their deficiencies to their infants during womb life, which begins a terrible dive to disease that can take the child's life or leave the child disabled.[308]

Health Restoration from Fossil Diet Diseases

Undernutrition and infectious diseases exist in a malevolent synergy. Malnutrition makes children biologically vulnerable. It degrades the immune capacity to defend against infectious disease, and disease further depletes and deprives the body of both energy and essential nutrients. The nutritional deficiency cycle amplifies poverty through lost wages, increased health care costs, and impaired intellectual development.

Stunting occurs in early development and cannot be undone. Economists estimate that stunting can reduce a country's GDP by as much as 12%.[309] Wasting occurs from a short-term lack of nutrition, such as a winter season. The child may survive but suffers from underdeveloped body and organs; brain, heart, eyes, lungs and musculature.[310] Freedom foods can relieve considerable burden of disease by supplying the full set of essential micronutrients to women of reproductive age and their children.

Premature births

Maternal micronutrients status before, during, and after pregnancy and during lactation is crucial for infant development. Mothers need the full set of micronutrients to pass on to their infants to build their brains, organs and neurological functioning. Micronutrients allow infants to build healthy brains, hearts, endocrine and immune systems.

Healthy newborns experience significantly fewer health issues at every life stage. Preterm births result in newborns that are extremely susceptible to major health issues throughout their life. Preterm births and a wide spectrum of development disorders for every major organ are associated with nutrient deficiencies that freedom foods resolve.

The 380,000 premature births in the US yearly is higher than any other wealthy country and resembles sub-Saharan Africa. Preterm birth is the most frequent cause of infant death, 17% of all infant deaths, and costs the US healthcare system $26 billion a year.[311]

Preterm birth occurs when a baby is born too early, before 37 weeks of pregnancy. In 2016, preterm birth affected about 1 of every 10 US infants. Preterm infants suffer low birth weight, stunting, cerebral palsy, intellectual disabilities, chronic lung disease, blindness and hearing loss. Development difficulties may result in problems breathing, feeding, seeing and hearing.[312]

146

One in six, or about 15%, of US children aged 3 through 17 years have a one or more developmental disabilities.[313] The mother's diet has an obvious impact on preterm birth. Women who eat a Western diet of high fat and sugar foods before they become pregnant are around 50% more likely to have a preterm birth than those on a healthy diet.[314] Women who eat a diet high in protein and fruit prior to becoming pregnant are significantly less likely to have a preterm birth.

Iron deficiency causes anemia, which is known to cause spontaneous preterm births. Women with iron-deficiency anemia that become pregnant have significantly lower energy and fail to gain adequate weight during pregnancy. With iron deficiency, the odds of low birth weight were tripled and preterm delivery more than doubled. When vaginal bleeding before entry to care accompanies anemia, the odds of a preterm delivery are increased fivefold.[315]

An obvious strategy might focus on specific micronutrient culprits for preterm births, underweight babies and infant development disabilities. Instead, freedom foods offer a smarter medical strategy – to insure mothers have the full set of nutrients they need, and their fetus needs, for healthy development. Research indicates that only tablespoon freedom food nutralence a day can resolve micronutrient deficiencies for mothers and infants.

Autism spectrum disorder

Iron deficiency in pregnant mothers appears to be a culprit in several diseases associated with early brain development, especially autism spectrum disorder and Asperger syndrome.[316] Children have a five times higher risk of developing autism if their mother had an iron deficiency, combined other risk factors such as a metabolic condition like obesity hypertension or diabetes.[317] Iron deficiency in a pregnant mother during womb-life predisposes the newborn to an under-developed brain and brain disorders.

Omega-3 in freedom foods or supplements have been shown in several studies to moderate the symptoms of autism spectrum disorder, ADHD and related brain maladies.[318] While algae compounds do not reverse autism – they essentially moderate the thunderstorm of disturbances in the neurological synapses and brain. This allows the patient to find calm.

147

Andrew Stoll at McLean Hospital studied omega-3 fatty acids in bipolar disorder.[319] Dr. Stoll found that patients who took omega-3 supplements had longer remissions between episodes of mood dysregulation. Joseph Hibbeln at the NIH has published several research studies and notes that "In the last century, Western diets have radically changed. People eat substantially fewer omega-3 fatty acids now and that rates of depression have increased by a hundred-fold." [320]

Captain Hibbeln and team were the first to establish a link among military personnel between low omega-3 levels and suicide risk.[321] Suicide risk was greatest among service members with the lowest levels of DHA, the major omega-3 fatty acid concentrated in the brain.

A recent review shows omega-3s are useful as add-on therapy in bipolar disorder too.[322] Since there is an increased prevalence of bipolar disorder in the extended families of autism patients, omega-3 has been proposed as a treatment for mood stabilization in patients with autism. Studies show that omega-3 can decrease hyperactivity children with ADHD.[323]

Cleft palate

Cleft palate remains one of the leading birth defects in the US. Micronutrient deficiencies, specifically folic acid and zinc during pregnancy, predisposes the development of cleft lip and palate. Obesity also increases the odds for an infant with a cleft palate. Nearly 7,000 babies born in the US each year suffer from a cleft palate and 4,440 babies have a cleft lip.[324] Children with a cleft palate have problems with feeding and speaking and are especially susceptible to ear infections. They also might have hearing and teeth problems. Surgical repair for a cleft palate and lip typically cost families $12,000 to $35,000.

A series of studies have concluded that other micronutrient deficiencies may increase the probability of preterm delivery. Research has identified Vitamins A, B and D, folate, and omerga-3 fatty acids.[325] A leading theory for some preterm births is attributable to the activation of the inflammatory pathway.[326] Several freedom foods block inflammation.

Allergies

The prevalence of allergic diseases such as asthma, atopic dermatitis, and allergic rhinitis has doubled during the last decade and contributed a great deal to morbidity and an appreciable mortality in the world. The onset of allergies in children has tripled in one generation.

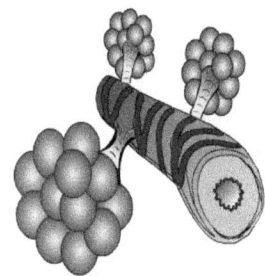

Over 40% of all US children now suffer from allergies.[327] Allergic diseases such as eczema, asthma and allergic rhinitis in infancy and childhood involve immune dysfunction. These hypersensitivity disorders involve strong inflammatory responses and the production of antibodies called IgE, which create asthma irritation, (left).

Seasonal allergies are severe in Japan, especially for pregnant women. The Osaka Maternal and Child Health Study examined the protective effect of the traditional Japanese diet of vegetables, fruit, antioxidants, fiber, and minerals compared with seaweed on allergic disorders. This cross-sectional study found that high dietary intake of seaweed, which is rich in calcium, magnesium, and phosphorus reduces the prevalence of allergic rhinitis.[328]

Algae medical tactics

Another promising line of research uses algae bioactive compounds to interrupt and stifle the allergy progression. Allergies are caused by an exaggerated reaction of the immune system to environmental substances, such as animal dander, house dust mites, foods, pollen, insects, and chemical agents. The event responsible for the development of allergy is the generation of allergen-specific CD4+ Th2 cells.[329]

Th2 cells produce a cascade of events that initiate the production of allergen-specific IgE, (immunoglobulin E) by B cells. Allergic reactions are induced upon binding of the allergen to IgE, which is tethered to the high affinity IgE receptor on the mast cell surface and basophils.[330] An allergy Cascade, below is adapted from *Thanh-Sang Vo, et al.*[331]

The algae inhibitors can block IgE cells from binding, inhibit histamine and cytokine production and release and suppress production of Th2 and IgE cells. These actions effectively disrupt the allergy progression, which gives relief to the patient.

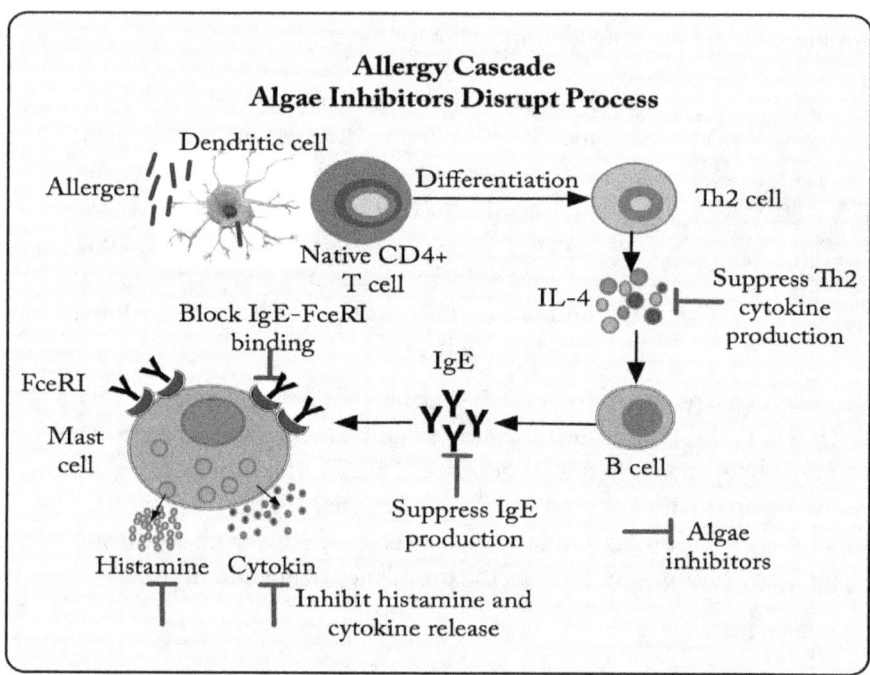

A UCSD team lead by Gregory and Mayfield found that certain algae-based proteins, Ara h 1 and Ara h 2, suppressed IgE antibody response to peanut allergy in mice.[332]

Current allergy drugs, such as antihistamines or corticosteroids, do ameliorate symptoms, but do not stop the disease progression. These drugs may cause serious side-effects, particularly in children and the elderly. Natural bioactive compounds from algae offer the development of new generation anti-allergic therapeutics, which can disrupt the disease cascade and impose fewer side-effects. Natural algae anti-allergy compounds have been used in Asian folk medicine for centuries.[333]

Many disorders follow a known sequence that builds one step at a time to create the chronic condition. Algae apply the disruption tactics, which have evolved over eons, to break the sequence and defeat the condition before it becomes chronic. The graphic illustrates that algae inhibitors apply multiple tactics to fight disease. They block, tackle, suppress, disrupt, gum and engulf. The "gum and engulf" strategies are not shown in the graphic, but algae compounds can gum up receptors, similar to the IgE cells, which prevent invasive cells from binding.

Algae lipids can the stimulate production of macrophages that engulf foreign substances, microbes, allergens, cancer cells, and anything else that does not have healthy compounds.[334]

Ecklonia cava, an edible marine brown alga, was tested for histamine release on human basophilic leukemia (KU812) and rat basophilic leukemia (RBL-2H3) cultured cell lines. Two algae bioactive phlorotannin derivatives suppressed IgE binding. The biomedical conclusion was that these algae derivatives are potential candidates for foods, cosmetics and pharmaceuticals.[335]

Obesity and diabetes

The CDC reported that one out of three American children born after the year 2000 will contract diabetes – predominantly due to a poor diet of nutrient-deficient calories.[336] Over 40% of women are likely to contract diabetes. The plague of obesity and diabetes creates havoc on our educational system and causes immense drag on our health system.

The cost of diabetes in the US exceeds $245 billion annually.[337] Neither the reported costs of obesity nor diabetes include the strain on education, social systems, businesses and the military. Our society will fail if we do not find solutions to obesity and diabetes – quickly.

Diabetes is a serious disease because it is associated with an increased risk of life-threatening complications such as a heart attack, stroke, or kidney disease. Overall, the risk for death among people with diabetes for these catastrophic complications is about four times that of people without diabetes. Diabetes carries with it significant risks for serious complications such as blindness, the need for dialysis, limb amputation and premature death. The cost of dialysis treatment for diabetics in the US exceeds $42 billion.[338]

Algae diabetes therapeutics

Diabetes mellitus occurs when blood sugar levels become elevated. Type 1 diabetes is associated with the destruction of the cells in the pancreas that manufacture insulin. Individuals with Type 1 diabetes require lifelong insulin for the control of blood sugar levels. In Type 2 diabetes insulin levels are typically elevated, indicating a loss of sensitivity to insulin by the cells of the body.

Research on humans and animals show algae components offer significant utility in the prevention and control of diabetes. Aligned studies have demonstrated algae's therapeutic value for the diseases common with diabetics; cholesterol management, blood pressure, heart disease and cancers.[339] Algae can moderate chronic inflammation that often precedes and accompany degenerative diseases.

Kelp, above, contain 13 times more calcium than milk and powerful antioxidants that are not found in land plants; fucoxanthin and fucoidan.[340] These macroalgae, above, are rich in B vitamins, C and K1, with high mineral content in magnesium, potassium and iron.

Appetite suppression

Global obesity levels have doubled in the last generation. The accumulation of excess body fat results when energy intake exceeds that expended. The impact on health has been profound since obesity is a major risk factor for most non-communicable disease including cardiovascular disease, cancers, chronic respiratory disease and diabetes.[341] Overweight and obesity adds further economic burden on already overstretched healthcare systems.

Energy balance is controlled by hypothalamic responses. Research shows that chronic overeating and obesity are due to elevations in endocannabinoid signaling.[342] The endocannabinoid system connects the brain to all peripheral organs, especially the stomach. Signals orchestrate food intake, energy balance, and reward.[343] It is comprised of lipid signaling molecules called endocannabinoids, which bind to cannabinoid receptors located on cells throughout the body.

A brain drug, Rimonabant, successfully blocks endocannabinoid signaling at cannabinoid receptors, which allows patients to reduce body weight.[344] The drug has limited availability in Europe. Trials have shown it causes such severe psychiatric side effects, it was not given FDA approval.[345] Algae bioactive compounds in functional foods can provide a natural solution to endocannabinoid signaling, without having to use synthetic drugs that try to fool the brain.

Nosh serves as an example. Nosh refers to eating greedily. Nosh has the connotation of eating on the sly, between meals, or possibly sweets in secret. Eating foods with empty calories leaves brain with no satiety, the feeling of fullness.

The old Cracker Jack box expressed nosh perfectly, "The more you eat, the more you want!" This was a successful tagline when most people were skinny. Borden's and then Frito-Lay had to change their advertising for modern obese children.

The problem with obesity in children is that children listen to their nosh signal and eat more food, more often. If the food contains empty calories, their little brains continue to signal: "I want more, more, more."

Health Restoration from Fossil Diet Diseases

A drug not approved in the US, Rimonabant, could help children, but the brain signal side effects are far too dangerous, except for morbid obesity.[346]

Algae compounds provide an array of medical benefits for children plagued with nosh signals that lead to obesity and diabetes. Two unique strategies may be called fill-gut and gut-full signaling.

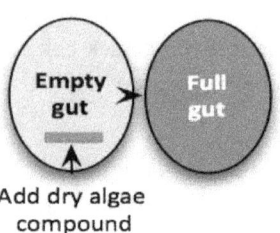

Add dry algae compound

The **fill-gut** strategy adds a few grams of dried algae eaten early in a normal meal, possibly as sprinkles on a salad or drink. Algae expand and fill the stomach. Alginates absorb 300 times their weight in water, which quickly fill the gut and suppresses appetite by sending the gut-full signal to the brain.

Gut-full signals work because algae compounds activate the stomach's natural satiety signals.[347] Satiety signals an immediate feeling of fullness, which tells the eater to stop eating naturally. Algae satiety signals the brain and quashes the nosh feeling.

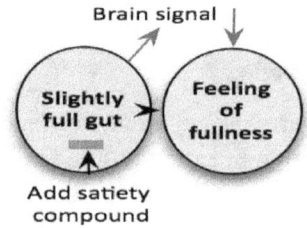

Add satiety compound

The nosh feeling hits children especially hard, which is why so many children are overweight. Many types of algae contain alginates that quash the nosh directive, and help young people stay at healthy weight.

Algae foods end nosh and create the gut-full signals

Strategies to prevent overeating

Sodium alginate reduces plasma glucose and protects the antioxidant system in diabetics.[348] Alginic acid and other compounds in sea vegetables and microalgae exert a protective effect against diabetes. Alginic acid improves cell sensitivity to the action of insulin, thereby improving glucose tolerance and normalizing blood sugar.[349]

Sodium alginate induces significantly lower postprandial rises in blood glucose, serum insulin and plasma C-peptides.[350] The addition of sodium alginate in the diet leads to a delayed gastric emptying rate induced by the fiber, which moderates glucose response. Algae polyphenol extracts have anti-diabetic effects through the modulation of glucose-induced oxidative stress.[351] These extracts slow starch-digestive enzymes such as alpha-amylase and alpha-glucosidase, which modulates the release of plasma glucose.

Do no harm

A thoughtful reminder from Hippocrates applies to algae therapeutics.

Make a habit of two things: to help; or at least to do no harm.
– Hippocrates

The National Institute of Health recommends algae compounds such as omega-3s because they are natural products and induce no side effects.[352] This natural "no harm" advantage makes them a first choice for autism spectrum disorder treatment over pharmaceutical drugs that impose significant side effects. The FDA certifies many algae-based freedom foods as GRAS, Generally Recognized as Safe because they do no harm.

Soluble fibers

Doctors recommend a high fiber diet for weight loss because fiber assists in shedding pounds, helps lower cholesterol, and reduces the risk of developing heart disease and diabetes.[353] Consumers need to be cautious with fibers because some fibers may cause an upset stomach. Many protein, fiber bars and drinks that are marketed as health foods are loaded with artificial sweeteners to improve taste, but they leave out essential vitamins, minerals and antioxidants. Instead of helping with weight loss, the added sugars contribute to weight gain. These bars and drinks are often packed with synthetic fiber from artificial sources. Artificial fibers create indigestion and stomach pain for many people.

All natural fibers are not created equal. Soluble fiber dissolves in water and insoluble fiber does not. Soluble fiber, found in algae, oatmeal, lentils, fruits, nuts and many beans helps lower cholesterol.[354] Foods containing soluble fiber benefit from water or other liquids. Consumers need to drink lots of water in order to reap the nutritional benefits. Soluble fiber also slows digestion, giving a feeling of fullness.

Insoluble fiber, found in whole grains, wheat bran, seeds, nuts, barley, and dark leafy vegetables, speeds up the digestion. Insoluble fiber is not absorbed by water and passes through the body quickly. Both soluble and insoluble fibers are useful for a healthy diet.

The plentiful soluble dietary fibers in algae help avoid obesity and diabetes. The total fiber content of several algae species, (6 g/100g), is more than double the of fruits and vegetables promoted for their fiber content: prunes (2.4 g), cabbage (2.9 g), apples (2.0 g), and brown rice (3.8 g).[355] Macroalgae, sea vegetables, are particularly rich in dietary fibers, right.

Seaweeds evolved with strong fibers so they could withstand the pounding of the surf where they grow in the intertidal zone.

Sea vegetables have a total dietary fiber content varying between 30% and 72% on a dry weight basis. The majority of algae fibers are water-soluble. Some species have 15% to 50% insoluble fibers, which can be helpful for certain dietary needs.

Algae fibers such as alginates are commonly used to make foods more attractive and more palatable. Algae offer an excellent source of fibers presenting chemical, physic-chemical, and rheological diversities that are beneficial in nutrition. Rheological characteristics define the stability and appearance of foods, *e.g.*, creaminess, juiciness, smoothness, brittleness, tenderness, and hardness.[356] Algae components such as alginates are used to improve rheological characteristics with emulsions, spreads and pastes.

Other research shows that viscous dietary algae fiber, moderates insulin resistance syndrome and other diabetes and coronary heart disease risk factors.[357] The properties of alginate solutions and gels suggest a suite of biomedical and pharmaceutical uses. No alginate functional food or medical products are yet on the market, but many foods and medical products contain alginates.

Heavy metals poisoning

Global warming amplifies water shortages that force communities to dig deeper wells. Deeper wells tend to increase levels of heavy metal poisons, especially iron, lead, mercury and arsenic. Heavy metal poisoning and arsenic in particular, right, causes horrific medical problems.

Eating algae-based foods, allows the body to bioabsorb the tiny algae cells that chelate with heavy metals. The body sluffs them off and passes them out of the body in the urine. In regions plagued with heavy metals in drinking water, (e.g. Bangladesh, China and communities near mines or power plants), the ability of algae to detox heavy metal poisoning can save millions from painful and ugly disability or death.[358]

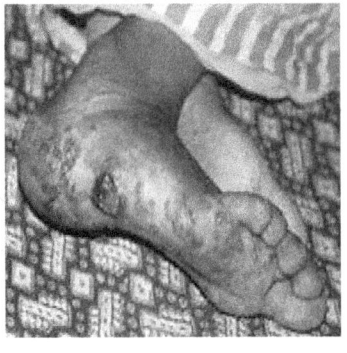

Pesticide pollution imposes a terrible toll on pregnant mothers, fetuses and young children.[359] Recent research shows a connection between pesticide exposure by pregnant mothers and autism spectrum disorder and other neurological diseases in children. Pesticides cause serious health impacts such as headaches and nausea and chronic impacts like cancer, reproductive harm, and endocrine disruption. People exposed to these poisons in crop fields, in the drinking water of local communities, and pesticide residuals on produce suffer terrible health risks. People may experience acute nerve, skin, and eye irritation, nerve or eye damage, headaches, dizziness, nausea, fatigue, and systemic poisoning.

Additional research suggests that the microcrop spirulina eaten normally can chelate with poison molecules, similar to their attachment to heavy metals, and flush the poisons from the body.[360]

Omega-3 fatty acids

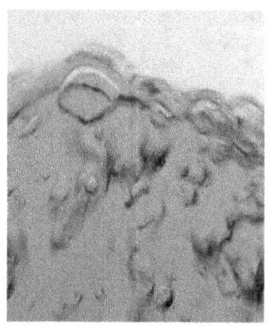

Omega-3 fatty acids, (left) found in algae manage cellular membranes and moderate inflammatory allergy responses. Omega-3s do not cure inflammatory conditions but can bring substantial symptom relief. High proportions of DHA and EPA in maternal and infant plasma phospholipids are associated with less IgE-associated disease and a reduced severity of the allergic phenotype.[361]

Studies show diet supplementation with omega-3 fatty acids, Zn and vitamin C significantly improves the asthma control test, pulmonary function test and pulmonary inflammatory markers in children with persistent bronchial asthma.[362]

Research results on EPA supplements alone have been mixed, possibly due to a high background intake of omega-6 fatty acids, the kind found in most vegetable oils. Medical research supports a diet with increased omega-3 fatty acids and reduced omega-6 fatty acids to protect children against symptoms of asthma and a litany of other autoimmune disorders.

The recommended intake ratio of omega-6 fats to omega-3 fats is 2:1 to 4:1. Americans tend to eat a ratio of 14:1 to 25:1 of omega-6 vs. omega-3 fatty acids.[363] High omega-6 can be harmful since omega-6 fats have inflammatory tendencies.

Other research indicates supplements rich in EPA and DHA reduce asthmas symptoms compared to children who took placebo.[364] Other studies found evidence for a modulatory effect of the dietary omega-6: omega-3 fatty acid ratio on the presence of asthma in children. These studies provide evidence that a diet with increased omega-3 and reduced omega-6 protects children against asthma and can treat symptoms.[365]

Exercise-induced asthma may cause bronchoconstriction from airway inflammation following exercise. A diet rich in pro-inflammatory omega-6 fatty acids (often associated with terrestrial crops), and low in anti-inflammatory omega-3 fatty acids tends to **amplify airway constriction.**

Researchers sampled mucus taken from participants found omega-3 supplements—3.2 grams of EPA and 2 grams of DHA daily—resulted in reduced amounts of pro-inflammatory cells and markers.

An Indiana University team found adults with persistent asthma that took an omega-3 supplement daily for three weeks improved their post-exercise lung function by 64%.[366] Improvements allowed a 31% decrease in the use of inhalers.

Benefits for athletes

Omega-3 supplements are popular for athletes and non-athletes for their ability to improve endothelial (i.e. blood vessel) function, reduce inflammation, and increase the provision of energy from fat.[367] Protein is an essential nutritional component in the required to repair and build muscle tissue broken down during exercise.[368] The FDA recommends between 1.2 and 1.7 g protein per kg body weight.[369]

Algae are rich sources of protein and contain all of the essential amino acids at various concentrations.[370] Algae provide a valuable resource for athletes requiring high levels of protein, especially for vegan athletes for whom eggs and dairy whey protein may not be suitable. The Cleveland Clinic's Dr. Patton notes omega-3 battles inflammation.[371] Exercise is a form of good stress on the body but results in the production of inflammatory free radicals that cause oxidative stress and can damage cells. Omega-3s assist by counteracting inflammation and reducing joint pain and tenderness associated exercise-induced inflammation.[372]

Omega-3s decrease delayed-onset muscle soreness, increases the rate of recovery, and reduces the risk for infection due to immunodeficiency.[373] They coat and protect cell membranes. When incorporated in the membranes of red blood cells, they increase the deformability of the red blood cells. This allows the cells to move swiftly through capillary beds and efficiently deliver oxygen and remove carbon dioxide.

Whole algae

The microalgae Spirulina and Chlorella are sold today as functional foods due to their extraordinarily high nutralence. They are generally regarded as safe (GRAS), by the European Food Safety Authority (EFSA) and by the FDA.[374] *Arthrospira platensis* spirulina is a filamentous cyanobacterium that has the highest recorded protein content (66%) of any whole food.[375]

Chlorella enjoys global sales exceeding $38 B. The main substance found in Chlorella that is beneficial for human health is β-1,3-glucan, which is an active immunostimulant, free-radical scavenger and reducer of blood lipids.[376] Chlorella is also rich in protein, (48% dw), polyunsaturated fatty acids, (39% of total lipids), and phosphorus. The nano-cells also include vitamins, (B-complex and ascorbic acid), minerals (potassium, sodium, magnesium, iron, and calcium), beta-carotene, chlorophyll, and Chlorella growth factor (CGF). *Chlorella's* antitumor polypeptide, (CPAP) has immunosuppressive, anti-inflammatory, anti-hypertensive, anti-atherosclerosis and antioxidant capacities.[377] Blue-green algae, cyanobacteria such as Spirulina have no cell walls, which makes the protein and nutrients readily bioavailable to the consumer. Many algae species evolved strong fibrous cell walls made of anionic polysaccharides to manage osmotic pressure within the cell. The cell wall makes digestion and protein extraction difficult for people.

Algae ingredients have been incorporated into a number of functional foods, including noodles, bread, biscuits, drinks, sweets, and beer. Several businesses have been set up for the sale of algal products, such as AlgaVia® which produces protein and lipid-rich Chlorella flour.[378] AlgaVia® (left), is a lipid-rich whole algae powder.

AlgaVia delivers a unique set of sensory and nutritional benefits to bakery products.[379] The powder reduces overall fat, saturated fat, cholesterol and calories. The whole food ingredient adds fiber and protein, is non-GMO, vegan, and gluten-free.

Food scientists have developed a wide array of treatments to disrupt the cellulosic cell wall with heat, ultrasound or other techniques that make algae proteins and other cell components more bioaccessible and easier to digest.[380] Ongoing research on the best pretreatment of algae cell material includes fermentation and various approaches to drying.[381]

HIV/AIDS

Algae lack an immune system. These tiny cells have developed an elaborate chemical defense strategy of bioactive compounds for protection against bacteria, viruses, and fungi, which pose significant threats to algae in natural settings. Algae evolved in ancient oceans surrounded by predators and pathogens. Incredibly, a milliliter of seawater may contain 10^7 viruses, 10^6 bacteria, and 10^3 fungal cells.[382]

Kubanek and her colleagues have identified more than 20 different antimicrobial compounds from the surfaces of two common macroalgae. These compounds have been shown to be effective against infectious and pathogenic microbes and show promise against human viral diseases.[383]

Lectins and phycobiliproteins are two families of bioactive algae proteins that have many food and medical applications. Lectins are most commonly extracted from macroalgae, while phycobiliproteins are typically isolated from microalgae.

Lectins bind with carbohydrates without causing modification with enzymatic activity. Lectins are involved in several biological processes, including host-pathogen interactions, cell–cell communication, induction of apoptosis, cancer metastasis and antiviral activities.[384] Due to their carbohydrate binding capacity with high specificity, lectins are used in blood grouping, anti-viral (including human immunodeficiency virus type 1, cancer biomarkers, and targets for drug delivery.[385]

Mikinori Ueno and team examined the effects of various acidic polysaccharides isolated from marine algae on the infection and replication of human immunodeficiency virus type-1 (HIV-1), hepatitis B virus, hepatitis C, and human T-cell leukemia virus type-1. The team found the sulfated fucoidan polysaccharide, ascophyllan, and two other fucoidans significantly inhibited virus infection.[386]

A Vietnamese team examined the anti-HIV activity of three fucoidans, extracted from three brown seaweeds. Fucoidans inhibited HIV-1 infection when they were pre-incubated with the virus. Each of the fucoidans blocked the early steps of HIV entry into target cells. Fucoidans were not effective after HIV infection.[387]

Some medical scientists recommend a diet that includes algae to protect from virus acquisition and to increase an immune response.[388] Epidemiology studies have shown dietary algae decreases HIV/AIDS viral fusion/entry and replication and increases immune response.

Regular algae consumption by Asian people could be an explanatory factor in their relatively low HIV/AIDS rates. HIV/AIDS rates in countries that regularly eat seaweed are extremely low, (<0.1%) compared to rates in the US, (0.6%).[389] Freedom foods can provide viral infection protection as well as increased immune response.

Summary

Freedom foods can resolve possibly 80% of the painful preterm deliveries and impaired infant development by providing the full set of healthy nutrients for pregnant women and infants. Freedom foods can reduce by possibly 60%, other child development maladies, including autism and cleft palate.

Nutrient deficiencies are multifactorial and consequently difficult to pin to specific conditions with precision. However, new tools such as big data, smart analytics, machine learning, IoT and artificial intelligence are teasing out critical associations among micronutrient deficiencies.

The next chapter explores microcrop therapeutic solutions.

13. Health Restoration with Microcrop Therapeutics

The development and conveyance of microcrop medicines represents the most credible action to support microcrop food adoption by mainstream consumers. Soon, the artificial boundary between food and medicines will disappear.

Key takeaways

- What health maladies do microcrop therapeutics address?
- What are environmental issues such as nitrogen?
- Is intensive mechanical agriculture sustainable?

Roughly 70% of medicines currently come from terrestrial plants or animals. Many of those will be replaced because microcrop can produce the same or better therapeutic compounds substantially faster, cleaner and at a higher quality. Consumers will delight in their new freedom to choose natural compounds that do not impose the high financial costs and heavy drag of undesirable drug side effects that are common today.

Today, people tend to think of food, pharmaceuticals and medicines separately because they are marketed in different stores. Tomorrow, those lines will blur. Amazon, Target, Walmart, Kroger and others will build supply chains that deliver food, medicines, pharmaceuticals and cosmeceuticals. Consumers will have the freedom to choose foods that prevent illnesses and other foods that treat specific diseases.

Market research shows that people strongly prefer following the advice of Hippocrates:

Let food be thy medicine in the medicine be thy food.

Health Restoration with Microcrop Therapeutics

Algae already support our global food system as useful ingredients, valuable compounds for functional foods, and in some cases, algae food. Tomorrow consumers will choose algae foods for their extensive medical benefits as well as freedom from allergens and empty calories.

Normal health is not the place a smart person wants to be when the CDC reports that 71% of the US population is overweight or obese, and over half of adults over 50 have one or more chronic diseases.[390]

Microcrop medical treatments

Microcrops such as algae offer therapeutic protection or treatment for many diseases. A web search of algae and any disease typically turns up relevant recent research. Peer-reviewed scientific research provides algae-based medical solutions for the following health challenges.

Deficiencies	Major organs	Major systems	Diseases
Vitamins	Brain	Cardiovascular	Blood pressure
Minerals	Eyes	Digestive	Hyperlipidemia
Elements	Heart	Endocrine	Bleeding gums
Antioxidants	Lungs	Immune	Infections
Hormones	Kidneys	Respiratory	Inflammation
Disorders	Skin, hair, nails	Circulatory	Cancers
Mood	Liver	Urinary	Immune
Anxiety	Blood	Nervous	Viral infection
Psychotic	Pancreas	Muscular	Bacterial infection
Personality	Hypothalamus	Integumentary	Cuts and burns
Sexual	Pituitary	Reproductive	Diarrhea
Development	Thyroid	Skeletal	Diabetes
Brain	Nerves	Lymphatic	Obesity

Algae prevent or remediate nutrient deficiencies, which inflict pain and development disorders on over half of the world's people. Algae compounds offer medical treatments to restore healthy functioning to major organs and to major body systems.

Algae have therapeutic properties that improve health and prevent disease that are not found in terrestrial plants. Algae evolved in incredibly harsh environments and developed hundreds of bioactive compounds that defend against predators and disease. Land plants invest most of their energy in roots, sexual apparatus and physical structure. Algae do not waste energy on those attributes, so the plant has more energy to invest in building bioactive compounds.

Algae bioactive compounds include anticancer, antiobesity, antidiabetic, antihypertensive, anti-hyperlipidemic, anti-coagulant, anti-inflammatory.[391] Other bioactive compounds have properties for immunomodulatory, anti-estrogenic, antiviral, thyroid stimulating, neuroprotective, antifungal, antibacterial and tissue healing.

Active compounds include sulphated polysaccharides, phlorotannins, carotenoids, (e.g. fucoxanthin), minerals, peptides, and sulfolipids, with benefits against the generative metabolic disorders.[392]

Metabolic syndrome

A cluster of conditions causes metabolic syndrome — increased blood pressure, high blood sugar, excess body fat around the waist, and abnormal cholesterol or triglyceride levels.

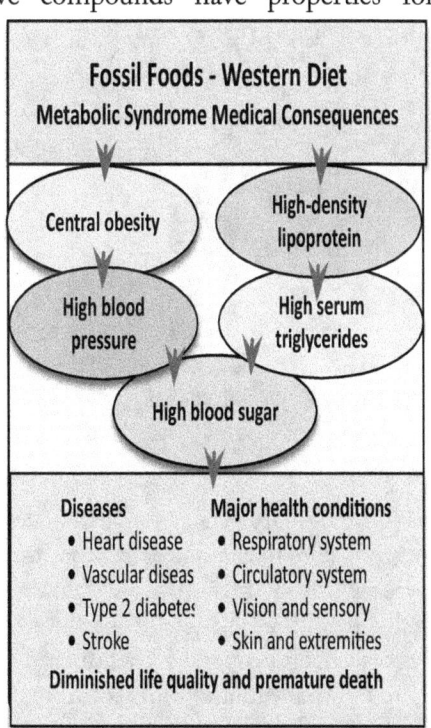

Fossil Foods - Western Diet
Metabolic Syndrome Medical Consequences

Central obesity

High-density lipoprotein

High blood pressure

High serum triglycerides

High blood sugar

Diseases	Major health conditions
• Heart disease	• Respiratory system
• Vascular diseas	• Circulatory system
• Type 2 diabetes	• Vision and sensory
• Stroke	• Skin and extremities

Diminished life quality and premature death

These tend to occur together, increasing risk of the common diseases. The incidence of the metabolic syndrome steadily increases worldwide.

The notable exceptions are several Asian countries where seaweeds, macroalgae are consumed. Several studies show that diets that include only 6 g/d of sea vegetables improve the critical inflammation biomarkers and blood lipids.[393] Sodium alginate from sea vegetables, is used in culinary physics at some of the best restaurants in the world.[394]

Molecular gastronomy investigates the physical and chemical ingredient transformations that occur while cooking.[395] The study includes the social, artistic and technical components of culinary and gastronomic phenomena such as spherification of juices and other liquids. Chefs combine sodium alginate with calcium lactate to create spheres of liquid surrounded by a thin jelly membrane. High-end restaurants present these spheres (left), with different internal liquids for cocktails, appetizers, side dishes and desserts.[396]

Alginic acids are made up of hydrophilic colloidal polysaccharides that deliver complex molecules such as peptides, proteins, nucleic acids, oligonucleotides, and plasmids across biological surfaces. The delivery capability makes alginic acids ideal carriers for obesity, diabetic and related medicines. Many weight loss medicines use alginate as an appetite suppressant.[397] The mechanism includes the absorption of water to create the feeling of satiety or fullness. Researchers at Newcastle University found that dietary alginates can reduce human fat uptake by more than 75%.[398]

Sodium alginate acts as a natural chelator for metals. The nutraceutical industry uses it as a detoxifier that removes heavy metals from the body. Research shows sodium alginate pulls heavy metals including radioactive toxins from the body, such as iodine-131 and strontium-90.[399] Spirulina has also been proven to chelate and remove both heavy metals such as mercury and lead and radioactive toxins from human tissues.[400]

The effect of soluble fiber on the blood glucose response seems related to its ability to increase the viscosity of a meal.[401] Viscous fibers slow the gastric emptying rate of a meal in subjects with and without diabetes. Alginate fibers offer a source of viscous dietary fiber in algae-based foods. The main constituents of alginates are uronic acids (mannuronic and guluronic acids), which give the alginate characteristics similar to pectin (galacturonic acid).[402]

Hypertension – blood pressure

The CDC reports that high blood pressure affects over 75 million Americans and adds $94 billion a year in medical costs and lost work days.[403] Progression of hypertension results in cardiac and vascular abnormalities, such as endothelial dysfunction, peripheral resistance, altered contractility, and vascular remodeling.

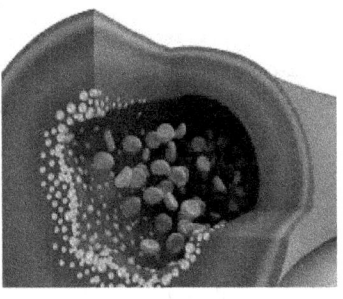

Hypertension increases the force of the blood against artery walls and eventually causes health problems, such as heart disease, strokes, kidney damage, organ damage and heart failure. High blood pressure is called the "silent killer."

Omega-3 supplements have been shown to lower total blood fat, which lowers blood pressure.[404] DHA and EPA supplements can also cut elevated triglyceride levels.[405] Omega-3 fatty acids also appear to lower the overall risk of death from heart disease. Omega-3 may reduce arrhythmias and slow the development of plaques in the arteries. Omega-3s moderate inflammation and help sustain tissue pliability. Patients who take omegae-3 supplements after a heart attack cut their risk of having another heart attack significantly.[406]

Ben-Gurion University researchers in Israel isolated a microalgal strain, which can accumulate up to 15% dry weight of a polyunsaturated fatty acid, PUFA called DGLA (Dihomo-γ-Linolenic Acid).[407]

Health Restoration with Microcrop Therapeutics

Andy Ayers, ex-CEO of Algae Biosciences in Arizona developed a natural algae mutant strain that produced over 30% dry weight of EPA oil.[408] Some algae strains produce predominately DHA, EPA or DGLA, while others produce a mix of two or more PUFAs.

Hyperlipidemia

Cholesterol and triglycerides are critical blood fats and lipids. Cholesterol is an essential component of cell membranes, brain and nerve cells, and bile, which helps the body absorb fats and fat-soluble vitamins. Cholesterol is a waxy substance that can build up as plaque on blood vessels. Foods containing cholesterol, saturated fat, and trans fats can raise blood cholesterol levels, such as dairy products, fried and processed foods and red meat. Lowering harmful cholesterol levels reduces the risk of heart attack, stroke, and other problems.

The Mediterranean diet, rich in vegetable foods, contributes both quantitatively and qualitatively to essential fiber compounds that decrease lipids and lower blood sugar levels. These include cellulose, hemicellulose, gums, mucilages, pectins, oligosaccharides, lignins. Scientists believe that dietary algae fibers may make it possible to lower dosages of hypolipemic drugs, which decreases possible side effects.[409] Microalgae produce bioactive metabolites that can be used in managing hyperlipidemia, microbial infection, and oxidative stress.[410]

Several studies have demonstrated that peptides derived from algae proteins possess antioxidative and antihypertensive properties.[411] Elevated levels of reactive oxygen species can cause oxidation of biological macromolecules, ultimately leading to pathological conditions that include endothelial dysfunction and hypertension. Oxidative stress is not only a causative factor of hypertension but acts as an important mediator in the imbalance between vasoconstrictor and vasodilator mechanisms. Oxidative stress-mediated events are basic factors in the development of hypertension, which can be moderated by algae antioxidant peptides.[412]

Other lines of research demonstrate algae's ability to decrease lipids and lower blood sugar, which improve cardiac and diabetic symptoms.[413]

Algae polysaccharides, such as carrageenan, are excellent sources of dietary fiber that moderate hypoglycemic events and lower cholesterol and lipids. Algae extracts produce low-density proteins in blood cholesterol, which helps regulate lipids.[414]

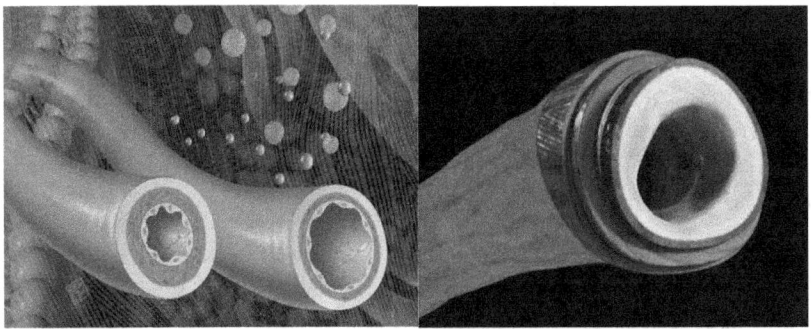

ACE inhibitors **relax blood vessels**

Alginic acid has been shown to exhibit antioxidant and angiotensin-converting enzyme, (ACE) inhibitory activities.[415] Alginic acid reduces tension on blood vessels, lowers blood flow and causes dilation of blood vessels, which results in lower blood pressure. ACE inhibitors are used to treat hypertension, cardiac failure, diabetic nephropathy and renal failure. These soluble polysaccharides act as prebiotics, stimulating growth of beneficial bacteria in the colon.

Anticoagulants

Anticoagulants are substances that prevent blood coagulation, (clotting).[416] These substances occur naturally in leeches, blood-sucking insects, and some algae species. Algae offer a broad array of novel sulfated polysaccharides with diverse chemical characteristics that produce a variety of natural anticoagulants useful for various medical conditions.

Anticoagulants and antiplatelet drugs reduce the risk of blood clots.[417] They are often called blood thinners, but these medications do not thin blood. They help prevent or break up dangerous blood clots that form in blood vessels or on the heart. Without treatment, these clots can block circulation and lead to a heart attack or stroke. They are often the first medication prescribed by doctors following a stroke.

By reducing the ability of the blood to clot — and thereby reducing the likelihood of coronary or vascular emboli — anticoagulants are frequently used in patients who are already at high-risk for stroke.[418] Unfortunately, these drugs can cause serious side effects, including blood in the urine or stool, nosebleeds, bruising, heavy bleeding, black vomit, dizziness, breathing difficulty, fever, chills, and severe chest and stomach pain. Natural anticoagulants from algae are likely to perform well, but do not induce unfavorable side effects.

Phlorotannins and sulfated polysaccharides, such as fucoidans in brown algae, carrageenans in red algae, and ulvans in green algae have been recognized as effective anticoagulant agents.[419]

Both freshwater and marine algae offer a large number of natural anticoagulant polysaccharides that have been isolated and characterized. Algae polysaccharides typically exert anticoagulant activity through antithrombin III and/or heparin cofactor II. These are important endogenous inhibitors, called SERPIN.[420]

The anticoagulant mechanism occurs when heparin, heparin sulfate and dermatan sulfate exert their activity. Some algae anticoagulant polysaccharides exert anticoagulant activity by directly inhibiting fibrin polymerization and/or thrombin activity. New functions of algae anticoagulant polysaccharides have been discovered recently.[421]

Heparin and its derivatives play important roles in many biological processes. It biostimulates the body to produce natural compounds that keep the blood flowing smoothly. Algae anticoagulant polysaccharides activate the fibrinolysis system and modulate endothelial cell functions.[422] Biologically active compounds in algae have displayed anti-platelet and anticoagulant proteins and fibrinolytic enzymes.

Hepatic inflammation

Human longevity rise has been accompanied by growing incidences of other devastating age-related pathologies. Aging is a progressive functional deterioration associated with frailty, disease, and death.[423] Stress and immune responses deteriorate during aging, causing low-grade inflammation and increased susceptibility to infections, which collectively lead to chronic disease. The liver is the second largest organ in the body and processes everything people eat and drink. It filters out harmful substances from the blood, and it is responsible for helping the body fight off infections.

Age-related hepatic, (liver) degradation occurs, such as an increased hepatocyte size, an increase in the number of binucleated cells, and a reduction in mitochondrial number.[424] These changes significantly affect liver morphology, physiology, and oxidative capacity. Aging predisposes patients to hepatic functional and structural impairment, inflammation and metabolic risk, which can cause non-alcoholic fatty liver disease, (NAFLD). No validated treatments of NAFLD exist beyond weight loss or comorbidity management.[425] Algae may provide a solution.[426]

Spirulina is one of the most important healing and prophylactic nutritional ingredients of the 21st century due to its nutrient profile, its therapeutic effects, and its lack of toxicity.[427] Spirulina may protect the body from hypertension, inflammatory diseases, insulin-resistance, diabetes mellitus, non-alcoholic fatty liver disease, malnutrition, anemia, allergic rhinitis, cancer, and reduction of drug toxicity.[428]

Emerging evidence shows extensive interaction between the gut microbiota, the immune system, and inflammatory pathways that influence aging.[429] The aging process can seriously affect the composition of the gut microbiota, which causes a common aging symptom; constipation. This uncomfortable indicator results from decreased intestinal motility and slower intestinal transit. The event chain reduces bacterial excretion, alters the gut fermentative processes and negatively affects the composition of intestinal microbiota. Counteracting inflammation that occurs in the liver and intestines through gut microbiota modulation, may be a factor for healthy aging.

Audrey Neyrinck and team at the Louvain Drug Research Institute in Belgium, found spirulina improves several immunological functions.[430] They assigned mice to three groups. Young three-month-old mice were fed with the standard diet, and old mice the standard diet supplemented with or without 5% spirulina for six weeks. The 5% addition of spirulina to the diet of old mice modulated immune function involving, among others, the TLR4 pathway.[431]

Oral consumption influenced both gut immunity and systemic sites, including the liver. This suggests that spirulina's immune action goes beyond the gut immune system. The team found improvement of the homeostasis in the gut ecosystem, which is essential for the gut health.[432] They concluded that spirulina improved gut microbiota in the elderly mice. It seems to be an effective dietary supplement for preserving a healthy gastrointestinal microbial community, in addition to its beneficial effects on immune function.

Cancer

Cancer is a complex set of diseases in which abnormal cells divide without control and invade other tissues. Cancer cells can spread to other parts of the body through the blood and lymph systems. Many of the more than 100 different types of cancer are named for the organ or type of cell in which they start. Cancer that begins in the colon is called colon cancer, while cancer that begins in melanocytes of the skin is called melanoma. In 2016, 1.7 million new cases of cancer were diagnosed in the US and about 600,000 people died from the disease.[433] Direct medical costs for cancer were about $125 billion in 2011. Death rates for the least educated are 2½ times higher than the most educated.

Medical tests with algae compounds have addressed more than 70 types of cancer.[434] Therapeutic strategies use algae compounds, such as astaxanthin and metabolites, including zonaquinone acetate and flabellinone that create growth inhibitory effects on cancer cells.[435] The natural algae compound fucoxanthin has strong antioxidant and cytotoxicity against breast, lung and prostate cancer. Algae sulfated polysaccharides have been shown to inhibit cell proliferation and to induce apoptosis by inhibiting IGF-IR signaling in several cancers. Other algae metabolites operate as antioxidants and protect against DNA damage.

Health Restoration with Microcrop Therapeutics

Medical scientists are working to develop novel targeted therapies in human cancer treatments against cancer tumors, skin melanomas, (below), and other forms of cancer.

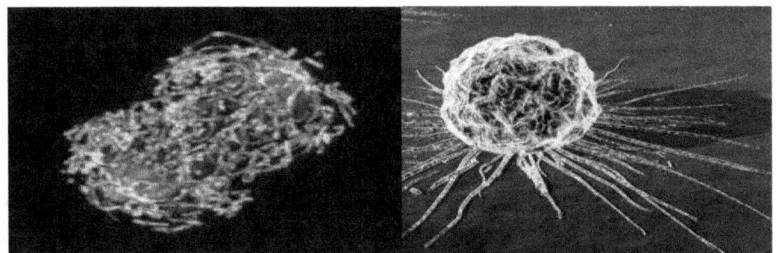

Targeted methods are capable of selectively killing cancer cells, but are not harmful to normal cells.[436] New anti-cancer therapies target molecular path-ways thought to be critical for tumor survival, and metastases. Some algae-derived compounds modulate multiple cellular mechanisms, including cellular cytotoxicity, inhibition of invasion and the eradication of cancer cells.[437]

These potent, naturally occurring anticancer compounds effectively prevent tumorigenesis. Fucoxanthin, a carotenoid found in diatoms and brown seaweeds exhibit anticancer activity, of cancer cells along with the induction of cancer suppressor genes and cell cycle arrest, but not apoptosis, (cell death).[438]

Sung-Suk Suh and team demonstrated with in vitro experiments that an extract from the Antarctic freshwater microalga, *Chloromonas sp.*, exhibited high anti-oxidant capacity and triggered anticancer activities, including anti-proliferation, anti-invasion and apoptotic cell death in cancer cells through the modulation of apoptosis-related genes.[439]

Metastasis, the leading cause of cancer mortality, occurs with a sequence of events, including invasion. In the invasion process, malignant tumor cells are able to become dissociated from the primary tumor and invade the surrounding tissue through the modulation of proteins involved in the control of cellular motility and migration. Algae extracts can significantly inhibit the invasion process of cancer cells at low doses, relative to those with anti-proliferative effect.

173

The algae extract in the study had sufficient antioxidant activity to induce apoptotic cell death in cancer cells *in vitro*, through the caspases dependent pathway.[440]

Algae also produce steroids that inhibit both cancer cells and cancerous tumors.[441] Scientists are discovering new bioactive steroids in algae species such as Sargassum that exhibit cytotoxic activity (toxicity) against various cancer cell lines.[442] A new promising line of research focuses on cancer vaccination, to make cells immune from the onset of cancer. One of the most successful approaches has been working with genetically modified tumor cells that are introduced into a patient's body. Algae toxins enable researchers to harvest natural toxins that strengthen the immune response, which is the critical path to human trials.[443]

No other recombinant protein production system can accumulate these complex eukaryotic toxin molecules as soluble and enzymatically active proteins.[444] These traits set algae apart from other expression platforms. The potential of immunotoxins as potent and specific anticancer therapeutics is enormous.[445] The use of antibody drug conjugates, using small-molecule drugs, to target and kill cancer cells and to minimize the exposure of healthy cells is already a reality. Some expensive therapies are in late-stage clinical trials or are already approved by the FDA.

Protein toxins are effective in inhibiting cancer-cell proliferation, but their production is limited to bacterial expression platforms that require the protein to be denatured and subsequently refolded. The process increases production time and cost substantially.

Algae provide a faster, less costly production platform that adds the ability to create more complex molecules than presently can be produced in bacterial systems. Although additional work needs to be done to determine whether larger or smaller immunotoxins are more effective for specific cancers, the αCD22PE40 and αCD22HCH23PE40 produced by the UCSD team demonstrated that *C. reinhardtii* chloroplasts can create complex immunotoxins and provide an effective platform to produce next-generation cancer therapeutics.[446]

Stephen Mayfield and the UCSD Laboratory team developed a novel therapeutic anti-cancer strategy, which may be called "search, find, and kill."[447] The team developed a genetically engineered designer alga for the production of special compounds for both searching for cancer cells and then killing them. The first step introduces an antibody that hunts down a cancer cell, and then lights it up. The algae toxin is sent to find the cell based on its heat signature. The toxin follows and kills the "hot" cancer cell, while leaving healthy cells uninjured.

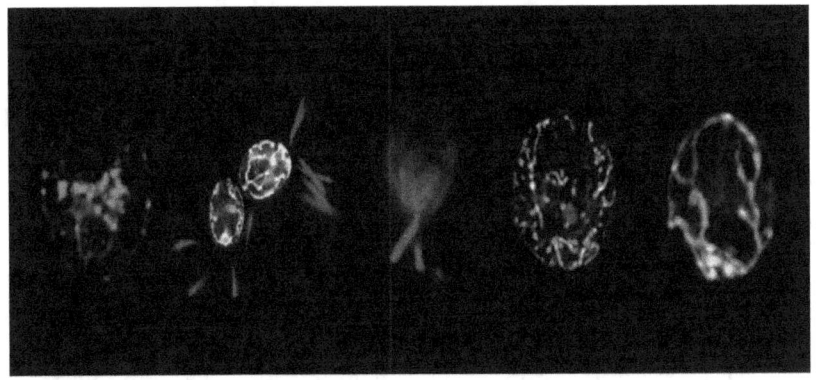

Stephen Mayfield's team studies beautiful therapeutic proteins

Mayfield calls these "dual-domain drugs," which offer better cancer therapy.[448] These drugs are currently produced through a complex process so costly that a course of treatments for lymphoma can cost over $100,000. The new algae-based production method should the cut treatment cost by possibly 90%.

The UCSD team demonstrated that the green algae *Chlamydomonas reinhardtii* are capable of expressing, folding, and accumulating a range of human therapeutic proteins in the chloroplast.[449] They also showed that recombinant proteins could be secreted from algae. Cost is a critical factor in the production of protein-based therapies. Algae reduce production costs because growing algae requires only fertilizer, trace minerals, and sunlight.

Algae have the potential to produce a wide variety of recombinant proteins for various therapeutic applications. An algae platform can produce desirable classes of therapeutically relevant proteins quickly. Algae offer the potential to produce a number of novel proteins due to the unique biochemical environment of the chloroplast.

Algae's unique ability to fold, assemble and accumulate multiple domain proteins as soluble molecules offers significant advantages.[450] The attributes that truly distinguish algae from other recombinant expression platforms are the presence of chloroplasts and the ability to produce and accumulate immunotoxin proteins in these compartments. Chloroplasts of higher plants such as tobacco could theoretically provide a viable option for expressing immunotoxins. Algae offer far better quality control because cultures can be grown in closed systems, avoiding problems from cross-contamination with native species.

Blindness

Over 15 million people suffer from some form of blindness, with the most common conditions being retinitis pigmentosa and age-related macular degeneration.[451] Both of these conditions occur when disease or age damages photoreceptors in the eye. Photoreceptors are responsible for transforming light entering the eye into electrical impulses, but when damaged, the brain is unable to receive this information.

Algae have 3.7 billion years of evolutionary experience with light. Since algae use light for energy, early algae evolved a gene that helped algae recognize the path toward light. Algae's ability to recognize light offers several lines of fascinating research that show promise for sight restoration. Retinitis pigmentosa is an eye disease in which there is damage to the retina.[452] The retina is the layer of tissue at the back of the inner eye that converts light images to nerve signals and sends them to the brain. Retinitis pigmentosa is a genetic disease that causes first tunnel vision, then night blindness and eventually blindness.[453] The disease impacts over two million people a year.

The disease degrades the outer retina's rod cells, which are highly sensitive to light and allow night and peripheral vision. Some patients lose only night vision. Others lose daylight vision and some go blind completely. The color-sensing cone cells of their inner retinas slowly degenerate. Although scientists do not clearly understand the pathway, the disease makes cone cells unresponsive to light but may not kill the cells. This phenomenon leaves a time window where the cone receptors are still there, but not functioning properly. Human solutions first had to recover sight for three blind mice.

A fascinating series of studies at Johns Hopkins, Harvard, and Stanford proved that inserting algae pigments into blind mice allowed the mice to first detect light, then see the direction of light, and finally see enough light to successfully navigate their way out of a maize.[454]

Neuroscientist Alan Horsager, at the Institute of Genetic Medicine at the University of Southern California is working with the gene responsible for making Channelrhodopsin-2, (ChR2) in algae.[455] This gene can detect light projected to the patients' eyes. This gene therapy is not yet approved for humans but is likely to offer applications for eye diseases like macular degeneration and retinal damage due to diabetes.

Human blindness

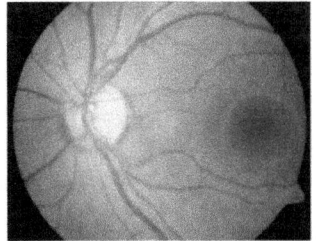

Algae are photosynthetic and very sensitive to light because photons provide their energy. Neuroscientists plan to use the same cells algae use to seek out sunlight for photosynthesis to replenish damaged cell equivalents in the human eye, (right).

Another approach uses optogenetics, which is a neuroscience technology designed to precisely control the activity of nerve cells.[456] It works by adding DNA instructions for channelrhodopsin, which algae use to sense sunlight and move toward it. Added to a nerve, it causes the cell to fire when exposed to a specific wavelength of light. The photosensitive protein helps direct algae toward a source of light.[457]

Health Restoration with Microcrop Therapeutics

The team led by E.G. Govorunova at Moscow State University, believe they can replace damaged cells in the retina with cells found in algae.[458]

Eye health products also continue to advance. Valensa International, recently received a patent covering its eye health cosmeceutical, EyePro MD.[459] The formula includes lutein, zeaxanthin and astaxanthin mixed with omega-3 essential oils from krill, algae or perilla seed. Lutein, zeaxanthin and astaxanthin can be extracted directly from algae oil.

Arthritis

Rheumatoid arthritis is one of many autoimmune diseases where PUFA supplements make a substantial difference. Most clinical studies examining omega-3 fatty acid supplements for arthritis have focused on rheumatoid arthritis (RA), an autoimmune disease that causes inflammation in the joints. Three meta-analyses of randomized controlled trials in RA patients found that omega-3 supplementation significantly decreased the number of painful and tender joints.[460] Several studies show that omega-3 supplementation creates decreases in pain intensity and shortens the duration of morning stiffness.[461] Omega-3s do not appear to slow progression of RA or to mitigate joint damage, only to treat the symptoms.

Diets rich in omega-3 fatty acids (and low in the inflammatory omega-6 fatty acids) may help people with osteoarthritis. Omega-3 supplements have been reported to reduce joint stiffness and pain, increase grip strength and improve walking pace for people suffering with osteoarthritis. Analysis of 17 controlled clinical trials examined the pain-relieving effects of omega-3 fatty acid supplements in people with RA or joint pain caused by inflammatory bowel disease and painful menstruation. The results show that omega-3 fatty acids, along with conventional therapies, help relieve joint pain occurrence and severity.[462]

Broader inflammation

Algae produce strong omega-3s, which can benefit people with a wide spectrum of inflammatory diseases.[463] Their high protein content provides a rich source of diverse biologically active peptides.[464] Select bioactive peptides provide potent inhibitory effects that moderate, disrupt or shut down the production of inflammatory mediators. Natural biologically active peptides protect cells by modulating the effects of oxidative stress.[465] Oxidative stress creates inflammatory reactions that cause endothelial failure (lining of blood vessels), lung disease, carcinogenesis, and atherosclerosis.

The versatility of algae solutions appears to derive from their high concentration of bioactive metabolites. Bioactive metabolites include brominated phenols and oxygen heterocyclics, nitrogen heterocyclics, kainic acids, guanidine derivatives, phenazine derivatives, amino acids and amines, sterols, sulfated polysaccharides, and prostaglandins.[466]

Fucoxanthin, a type of xanthophyll and an accessory pigment in algae chloroplasts, has also shown beneficial antioxidant, anti-inflammation, anti-cancer and anti-obesity activities.[467] Many algae species contain antioxidants such as astaxanthin, which scavenges free radicals and protects the body against oxidative damage. Astaxanthin moderates the lipids that raise LDL-cholesterol and strengthens cells and their membranes and tissues.

Edible vaccines

Infectious diseases account for more than 54% of total mortality in developing countries. Vaccines are the most effective means of prevention, but many are too expensive or too difficult to administer for broad-scale use. Expiration dates and refrigeration requirements inherent for nearly all commercial vaccines demand constant attention to the pathogen contained in such vaccines. This increases control, distribution and application costs. Vaccine degradation after stomach acid digestion and possible allergic reactions, also add constraints.

Edible vaccines offer an alternative method of vaccination that overcomes many of the disadvantages of current vaccines.[468] Edible vaccines do not require an extensive framework for their production, purification, sterilization, packaging, or distribution. This reduces costs up to 90% compared to traditional vaccines

Edible vaccines in functional foods provide nourishment, but their real value is their action to immunize the consumer against a disease. Upon

179

oral ingestion, the outer wall of alga cell protects the antigens from degradation by gastric secretion. The antigens are delivered to the intestinal mucosal surfaces, where they are absorbed by different mechanisms and stimulate a strong and specific immune response.

Edible vaccines are developed with an antigenic protein introduced into the algae cell by genetic engineering.[469] The antigen must elicit a strong specific immune response. The gene encoding for this antigen must be cloned into a transfer vector carrying an antibiotic-resistance gene. The vector carries the antigen into the algae cell. The edible malaria vaccine developed at UCSD provides a field-tested example.

Malaria vaccine

The parasitic, mosquito-borne, infectious disease malaria threatens nearly half of the global population. Many of the over 3 billion people most vulnerable to malaria are children and families that are extremely poor. Eradication of malaria requires low-cost, easily administered vaccines that work in concert with current control methods. Current medicines require refrigeration, which is difficult and expensive in hot rural areas. A short, low cost supply chain and ease of administration are essential components of a malaria vaccine, because malaria is endemic to regions that often lack an adequate healthcare infrastructure.

Mayfield's team collaborated with a medical team led by Joseph Vinetz in the UCSD School of Medicine, to create the precursor to a low-cost algae-based malaria vaccine that does not need refrigeration.[470] Recent work has focused on identifying specific parasite antigens that elicit the desired cellular and humoral immunity. These subunit vaccines are generally made in recombinant systems, purified, and delivered via injection. Production and purification of subunit vaccines are often complex and expensive. Bacteria, yeast, insect, and mammalian cells are most commonly used for producing recombinant proteins, but they are slow to produce and extremely expensive to extract.

James Gregory, Stephen Mayfield and their team developed an inexpensive malaria vaccine from plants and algae.[471] Their expression platform is capable of producing a recombinant antigen that precisely mimics the native protein structure.

Imitating nature ensures that the immune response confers protection to the corresponding pathogen. This is difficult to achieve because predicting whether a heterologous platform can replicate the three-dimensional structure of a foreign protein is nearly impossible, particularly for unique or structurally complex antigens. The authors describe mosquito stage vaccines, called transmission-blocking vaccines, (TBVs) which focus on antigens from sexual stage parasites.

Antibodies to several of these proteins block parasite sexual development when taken up with *Plasmodium* gametocytes during a mosquito blood meal. This prevents mosquito infection and subsequent transmission to the next human host. Antibodies raised in mice to TBV candidate antigens have successfully blocked transmission in both animal models and standard membrane feeding assays, but have not advanced beyond safety tests in human clinical trials.[472]

The malaria parasite life cycle includes a sequence of potential points for vaccine intervention. The mosquito introduces sporozoites into the bloodstream. They invade the liver and then are released into the circulatory system via the lungs. Thousands of sporozoites travel to the mosquito salivary glands when the oocysts burst. The parasite life cycle repeats after being transferred to a new human host via the mosquito.

The UCSD team produced Pfs25 and Pfs45/48 in the chloroplast of the green alga *Chlamydomonas reinhardtii*.[473] They demonstrated that alga-produced Pfs25 (CrPfs25) elicits TB antibodies. *C. reinhardtii* is an extensively researched single-celled eukaryotic alga that has only recently been exploited as a platform for producing recombinant proteins. Algae have been used to produce industrial enzymes, vaccine antigens, and complex immunotoxins on an academic scale. Depending on the desired posttranslational modifications, transgenes can be expressed from the nuclear or chloroplast genome. The ideal malaria vaccine must be extremely inexpensive, heat-stable and easily administered.

Currently, oral vaccines are available for polio, rotavirus, cholera, and typhoid, but these vaccines are based on attenuated or heat-killed pathogens. Novel strategies are necessary to overcome the obstacles that block orally available subunit vaccines, especially for pathogens like malaria that cannot easily be cultured and affect poor regions.

The UCSD Center for Algae Technology

CTB-Pfs25 was produced as a fusion protein in chloroplasts and orally delivered to mouse pups in freeze-dried whole cells. This strategy was not commercially effective, but the team learned insights helpful for a future oral malaria medicine.

Genetically modifying algae

Food consumers are concerned about GMO food ingredients. In Europe, GMO foods are not allowed. Algae are at the same time the most abundant and diverse, yet the smallest and simplest organisms. These characteristics make algae cells ideal for genetic modification, but preferably with a required GMO label. Interestingly, some transgenic engineering is allowed under the USDA Plant Protection Act and current EU regulations.[474]

Four genetic manipulation methods are allowed in the EU without classifying an organism as a GMO.

1. Introduce a gene from the same genus as the one being manipulated. In biological classification, a genus comes above species and below family.

2. Edit the location or presence of genes in the same organism. This is often called gene knockin or knockout.

3. Genetically cross breed, which creates hybrids like nearly all modern non-GMO foods.

4. Mutagenesis and selection, which intentionally nudges the organism to mutate using a stressor such as heat. The best mutants are then selected for the next round of mutagenesis.

Another path uses genetic tools to find the critical gene pathway for the expression of a desired trait. Big data computer analysis then determines which plant cells may be cross-bred. This combination approach results in natural hybrids on the order of 10,000 times faster.

Genetic manipulation is not the only road to discovery. Many medical and other valuable biocompounds can be found by bioprospecting, but the process requires time and patience. Texas A&M's Nano Biosystems Lab is developing a chip that can screen natural algae cells quickly to determine certain parameters, such as growth speed and oil content.[475]

New chip screens will enable algae diversity can be exploited as a unique source of bioactive compounds like carotenoids, fatty acids, sterols, mycosporine-like amino acids, phycobilins, polyketides, pectins, halogenated compounds, toxins, and others.

Algae are cost-effective and safe hosts for expressing a wide array of recombinant proteins, including human and animal therapeutics, as well as industrial enzymes.

Biomanufacturing

The ancient process of biomanufacturing uses living organisms as raw materials to convert into desirable bioproducts, such as beer, wine, cheese, and bread. Advances in biotechnology have expanded the scope of bioprocesses, enabling the use of gene editing for recombinant proteins and small molecules. Current transgenic techniques with terrestrial plants and animals shows merit, but suffer from high costs, low expression levels for complex proteins, and unstable cell lines.

Algae provide a superb biomanufacturing platform for the production of recombinant proteins and small molecules for a range of industries including bioenergy, biopharmaceuticals, biomaterials, nutraceuticals, agriculture, health, cosmetics and personal care. Biomanufacturing with algae allows low cost production, safety, metabolic diversity, and scalability.[476]

Recent studies have demonstrated algae's ability to express, fold, post-translationally modify, and secrete complex mammalian and other eukaryotic proteins.[477] Beth Rasala and Stephen Mayfield published an informative recombinant table showing various proteins, their function, expression host, genome and notable results.[478] Rasala and Mayfield note that recombinant proteins display promising results for the production of orally available vaccines and gut-active biologics, as well as complex unique anti-cancer therapeutics.

New gene editing tools like CRISPR, (Clustered Regularly Interspaced Short Palindromic Repeats), allow some genetic manipulation without requiring a GMO label. CRISPR allows researchers to target specific stretches of genetic code and edit DNA at precise locations, modifying select gene functions. The Broad Institute posts a good video illustrating how CRISPR works.[479] A novel approach to resolving blindness provides a good example.

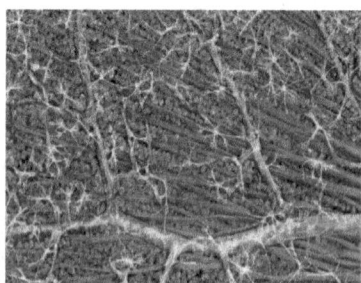 Retinitis pigmentosa is a group of inherited vision disorders caused by mutations in more than 60 genes. The mutations affect the eyes' photoreceptors, specialized cells in the retina (left) that sense and convert light images into electrical signals sent to the brain.

Health Restoration with Microcrop Therapeutics

There is no treatment for RP and disease progression results in blindness. A team led by Kang Zhang, Institute for Genomic Medicine at UCSD School of Medicine, used CRISPR/Cas9 to deactivate a master switch gene called *Nrl* and a downstream transcription factor called *Nr2e3* to restore sight in mice.[480] They reprogrammed mutated rod photoreceptors to become functioning cone photoreceptors, reversing cellular degeneration and restoring visual function in two mouse models with RP.

The major concern about outdoor production of GMO algae is that the cells may run amok. The fear is that the genetic modifications may invade natural habitats and change indigenous algae or other nearby microorganisms. Scientists at UCSD and Sapphire Energy successfully completed the first EPA-approved outdoor field trial of genetically engineered algae.[481] They found that cultivating the algae outdoors did no harm to nearby populations of wild algae.

The team tested the freshwater algae *Acutodesmus dimorphus* for 50 days in outdoor pools. The algae were genetically engineered by adding two genes; one for enhanced fatty acid biosynthesis, and one to make a green fluorescent protein. Unmodified algae from the same species were also grown. Neither the GMO nor unmodified algae escaped containment or were able to outcompete algae native to the area.

Summary

Microcrop medicines, with and without genetic modifications, will transform both disease prevention and treatment. Medical bioactive compounds can be found and grown in algae faster, at lower cost and with higher quality than terrestrial plants.

Individualized medicines will advance microcrop biofactories with compounds tailored to the DNA of each consumer. Development times for land plants take far too long for individualized medicine. Algae production platforms are able to produce bioactive compounds quickly and efficiently to support individualized medicine.

14. Eco Restoration – Air and Water

Imagine superior fossil-free foods that clean and restore our ecosystems rather than systemically extracting, wasting and polluting.

Key takeaways

- How do abundance methods work to restore degraded ecosystems?
- What are the mechanisms that allow environmental restoration?

The Emerald Renaissance and freedom foods health benefits to people and producers are substantial. While abundance methods preserve natural resources for future generations, they can also simultaneously restore health to degraded and destroyed ecosystems.

Emerald Renaissance - Ecological Restoration		
Air	**Water**	**Soil**
CCU - two tons of CO_2-e for every ton of algae	Produce 20% more fresh water than is used	Bring dead, abandoned soil back to life
Smoke stacks - coal, cement, industry	Clear pollutants from water	Restore the full set of micronutrients
Black soot particulates	Remove pharmaceuticals	Restore soil porosity
Biocycle animal carbon	Remove heavy metals	Recover from salt invasion
Smog - nitric, sulfur and toxic x-oxides	Capture and clean agri-overspill	Restore eroded and worn out soil
Direct air capture - CO_2	Eliminate water pollution	Restore soil biodiversity

Eco Restoration – Air and Water

Air

An air pollution solution can save millions from an agonizing premature death and simultaneously moderate global climate chaos. Freedom foods that capture and reuse carbon and black soot particulate matter, (PM) rather than releasing gases and particulates provides a positive strategy.

Air pollution is now the world's largest single environmental health risk. WHO reports that over 7 million people die prematurely annually. One in eight global deaths occurs from air pollution exposure.[482] Deaths occur from ischemic heart disease, strokes, lung cancer and respiratory infections. Air pollution costs countries over $1 trillion a year in health impacts and welfare losses.[483]

Combustion emissions in the US account for about 200,000 premature deaths per year due to elevated levels of PM fines, and about 10,000 deaths due to elevated in ozone concentrations.[484] PM premature deaths kill people about 10 years early, which means 200,000 premature deaths subtract 2 million life years from Americans. Over 90% people in cities are exposed to particulate matter fines in concentrations exceeding WHO air quality guidelines.[485]

Agricultural emissions are the leading cause of premature deaths from air pollution in the US, Europe, Russia, Turkey, Korea and Japan, followed by traffic and power generation emissions.[486] Ammonia and nitric oxides (right) volatizes from manure, fertilizers, and compost. These molecules mix with other air pollution sources such as traffic emissions, and have a huge impact on air quality.[487]

Nitric oxides, NO_x are particularly nasty irritant gases. At high concentrations, they cause airway inflammation and asthma. When the PM fines, chemicals, and pollutant gasses chemically react with water molecules and oxygen, they form acidic compounds. Acid compounds in acid rain or snow damage vegetation, buildings and the environment. Nitric oxide PM attract dust and create ozone and smog.

Carbon farming

> *Photosynthesis is simultaneously the cheapest and most*
> *efficient known solution of all nutrient cycling technologies.*

Air pollution from coal-fired power plants causes asthma, cancer, heart and lung ailments, neurological problems, acid rain, global warming, and other severe environmental and public health impacts. Besides the top line pollutants, CO_2, black soot PM fines, NO_x, SO_x and other harmful pollutants emitted by the US power plants annually include:[488]

- Over 41 tons of lead, 9,332 pounds of cadmium, 77,108 pounds of arsenic, mercury and other toxic heavy metals. Arsenic causes cancer in 1/100 people who drink water containing only 50 parts per billion.
- About 576,185 tons of carbon monoxide, which causes headaches and places additional stress on people with heart disease.
- Over 22,124 tons of volatile compounds, which form ozone.

Carbon farming with algae makes sense since algae are 200 times more efficient than trees in carbon capture. Carbon farming can occur with peace microfarms sited near GHG emission sources such as powerplants, cement and industrial plants. Small microfarms may use gas emissions from a brewery, restaurant or other carbon source.

David Dah-Wei Tsai and team at the Feng Chia University in Taiwan compared several algae cultivation systems with terrestrial plants for carbon uptake.[489] Carbon capture in the algae biosystems were close to 100%. The scientists found that trees and shrubs tested in 16 countries captured less than 20% of the available carbon.[490]

Algae offer the most promising solution for industrial capture of emitted CO_2.[491] Algae's ability to convert CO_2 into carbon-rich lipids and protein greatly exceeds food crops and does not compete for arable land or fresh water. Each ton of algae absorb 1.8 tons of CO_2. This exchange is possible because the atomic weight of C is 12, while the atomic weight of O is 16. Algae releases enormous amounts of O_2, 200 tons per hectare per year, while assimilating the C in its biomass.[492]

Microfarms can harvest carbon in the form of atmospheric or stack gasses. DOE has demonstrated that algae can capture 90% of powerplant CO_2 flue gas, 85% of NO_x and SO_x while producing 50 metric tons of algae biomass per acre per year.[493]

Algae eat their ancestors

Coal, shale, petroleum and natural gas are simply fossilized algae from ancient oceans. Algae are agnostic regarding their food (carbon) source. Algae can use the hydrocarbons in PM fines or coal dust, and organic hydrocarbons in wastewater. Combustion coproducts such as NO_x or SO_x also can be captured and effectively used as algae nutrients.

A multi-university consortium – Cornell, Duke and Hawaii – constructed a 7,000-acre integrated algae production with carbon capture system called ABECCS (algae bioenergy carbon capture and storage). The facility can yield as much protein as soybeans produced on the same land footprint, while simultaneously generating 17 million kilowatt hours of electricity and sequestering 30,000 tons of carbon dioxide per year. A portion of the captured CO_2 is used for growing algae and the remainder is sequestered. Biomass combustion supplies CO_2, heat, and electricity, thus increasing the range of sites suitable for algae cultivation.[494] While this project is not sustainable financially yet without a carbon trading exchange, it demonstrates potential for carbon farming.

Several power plant flue gas systems are under development. Michigan State University and PHYCO$_2$ has built a test carbon capture to algae process.[495] The University of Kentucky project with Duke Energy offers a good video. Global Algae Innovations in Hawaii has also posted a video of their robust carbon capture to algae production system.[496] Power Plant CCS provides a brief on algae carbon capture and storage.[497]

Obviously, algae cultivated with flue gasses or wastewater will contain nearly all the elements in those carbon sources, including toxic heavy metals. This limitation does not change the carbon farming calculus in harvesting carbon, but it does constrain bioproduct choices. One solution, biochar, sequesters carbon and heavy metals in a slow release fertilizer product that holds nutrients for 100 years in soils. Another, bioplastics and biodegradable building materials offer safe, long-term storage options.

Cow methane farming

Carbon farming can address animals too. One cow calf-pair create more CO_2-e carbon in methane than a car emits in a year. The insulated barn strategy, (planned but not yet implemented) will use algae construction-grade biofoam and biofiber materials to build barns that are nearly airtight. Dairy and meat animals raised in the barns will belch and fart their ruminant CH_4. The methane rises to a peaked roof where a smart sensor kicks-in a small electric motor to draw-in the methane. The gas, piped to a combustion chamber, drives a flywheel to produce energy, which is sent to an aluminum battery in the electric microgrid. The combustion chamber CO_2 exhaust bubbles into an algae biosystem where the carbon atoms are captured and converted to hydrocarbon biomass. This closed system makes the barn a net-zero carbon building.

Algae production can use flue or animal gasses but large microfarm sites will need more carbon or carbon sources are not available locally. Direct air capture, DAC provides a solution. DAC processes air directly from the atmospheric, removes CO_2 and purifies it. DAC may use a closed loop where the only major inputs are water and energy. The process creates an output stream of pure, compressed CO_2. Exxon-Mobile offers a good video of their approach.[498]

Direct air capture commercial scale reference design

Captured atmospheric CO_2 can be used to produce materials such as steel, concrete, fillers, and coatings, or to produce chemicals such as plastics, industrial chemicals, fertilizers, and carbonates. Microfarmers may bubble the CO_2 directly to feed their algae culture.

Carbon Engineering built a DAC demonstration facility that captures ~1 Mt-CO_2/year. The continuous process uses an aqueous KOH sorbent coupled to a calcium caustic recovery loop. The lead scientist, David Keith and team published a report concluding that their DAC technology levelized cost per ton CO_2 ranges from \$94 to \$232 per ton of carbon dioxide.[499]

Water

A literature search on algae and wastewater treatment returned 2,450 articles, where about a third focused on wastewater nutrients to biofuels. Wastewater treatment, the oldest algae application in the US, dates back before the Green Revolution began in 1950.

The illustration below shows a runway cultivation system where reclaimed water is a co-product.

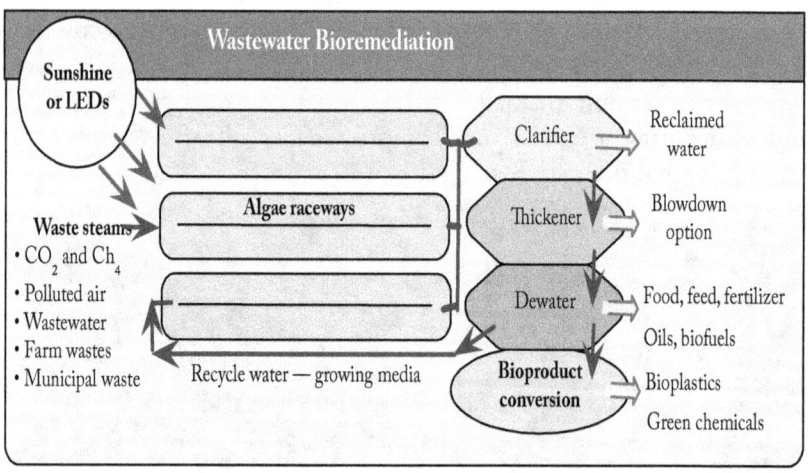

Algae biocycle nutrients in wastewater; clean water is one co-product

Algae have evolved to be the most important CO_2 fixers in aquatic ecosystems and the major biomass constituent in marine and freshwater environments. Scientists can achieve significant carbon capture by taking advantage of algae's unusual gift; to capture and fix CO_2 in wastewater.

Substantial research has demonstrated that algae can remove nutrients, toxic chemicals and pharmaceuticals from wastewater using any of several cultivation methods. Algae provide several benefits compared with conventional water-cleaning methods, such as activated sludge, including cost and speed. Algae can provide oxygen through photosynthesis needed for aerobic degradation of organic carbon and nitrification and harvested algae biomass can be used to produce high value chemicals or biogas. Algae wastewater treatment lowers pollutant chemical concentrations, including pharmaceuticals by > 90%.[500]

Algae biosystems can remove N from water. A study on closing nutrient cycles at the Netherlands Institute of Ecology demonstrates that algae can effectively recover nearly all the P and N from anaerobically treated black water, (toilet wastewater).[501] An algae biosystem captures the nutrients and generates algae biomass in one step. Screening experiments with green microalgae and cyanobacteria showed that all tested microalgae species successfully grew on anaerobically treated black water. A subsequent experiment in flat-panel photobioreactors, showed *chlorella* removed 100% of the P and N from the medium.[502]

Life cycle assessment indicates algae bioenergy achieves a substantially higher energy return than fossil fuels, which qualifies algae-based technologies as renewable. In 2019, the price for algae bio-oil is about twice fossil fuels. If algae were given half the subsidies benefiting Big Oil, algae biofuels would be cheaper than petroleum products. Some biotechnology scientists predict that algae liquid transportation fuels such as aviation gas will dominate 75% of the market.[503]

The water hungry textile industry consumes 2,700 liters (713 gallons) to produce the cotton needed to make a single t-shirt.[504] Textile processes consume half as much water again in pre-treatment, dyeing, printing, and finishing. The textile dyeing (right) and finishing industry has created a mammoth pollution problem. After agriculture, textiles are the No. 1 polluter of clean water. Over 3600 individual textile dyes are manufactured by the industry.

Textiles use more than 8000 chemicals in various processes of textile manufacture including dyeing and printing.[505] Textile processing, dyeing and printing consume massive quantities of water. An average textile mill producing 8000 kg of fabric per day consumes about 1.6 million liters of fresh water.

Algae and classical water treatment methods have particular difficulties with tanneries and textile wastewater. Textile industry uses many kinds of synthetic dyes, but the fabric uptake of these dyes is very poor. Textile plants discharge large amounts of colored wastewater heavily laden with chemicals. Colored textile wastewater severely impedes photosynthetic function. It also has an impact on aquatic life due to low light penetration, oxygen consumption and surface films. It may also be lethal to marine life due component metals and chlorine in synthetic dyes. Textile water treatment requires more sophisticated biotechnologies than other industrial sources. Several models appear promising, especially hybrid indoor / outdoor production.[506]

Frank Rogalla has developed a wastewater-to-biogas plant in Spain.[507] The facility grows algae in saline wastewater and harvests the algae oil. His team worked out the metrics and concluded the process achieves a positive energy balance. This energy and cost efficiency make algae-waste-to-energy a commercial viability.

Frank Rogalla and his team at Aqualia produced the algae oil used for the first Volkswagen powered by algae biogas. The All-gas, (algae in Spanish), waste water treatment plant located in El Torno, Chiclana, Spain. The vehicle's efficient hybrid engine generates zero emissions.

Freedom foods that biocycle nutrients and eliminate waste may be the most important new food technology this century.

15. Eco Restoration – Soil

A nation that destroys its soil destroys itself.
– Franklin D. Roosevelt

The soil is the great connector of lives, the source and
destination of all. Without proper care for soil we can have
no community. Without proper care for it we can have no
life.   *— Wendell Berry*

Key takeaways

- How do abundance methods work to restore degraded soil?
- How can biological systems enhance yields and restore fertility?

The elimination of monocultures, heavy equipment, cultivation, irrigation, chemical fertilizers, pesticides and other agri-poisons will eliminate soil structure degradation. Freedom foods include algae biofertilizer production of that can repair soil structure and restore fertility to degraded and even dead and abandoned cropland.

Algae biofertilizers have demonstrated their unique talent to restore degraded cropland, including the ability to bring dead soil back to life. Abundance methods that avoid chemical fertilizers and pesticides also protect local ecosystems. Freedom food growers can restore ecosystem health and biodiversity for future generations.

A project with Del Monte Fresh Produce division in Yuma Arizona shows the impact of algae biofertilizer on 200-acres of desert cropland that was exhausted, nutrient poor, compacted and abandoned.

SmartCultures, (Sustainable MicroAlgae Regenerative Technologies), delivered whole algae cells with the full set of nutrients to the melon crop via the irrigation in a drip system. SmartCultures are a derivative peace microfarm application designed to improve crop yields and soils.

Algae biofertilizer improved melon yields by >30%, compared with the control field and reduced inorganic fertilizer 55%, water 24% and total per-acre cost by 22%.[508] Soil porosity, looseness, improved 500%, which allowed roots to grow deeper for nutrients and moisture. Porosity made room for many additions to the symbiotic microbial community.

Algae biofertilizer regenerates soil microbes and fertility

IMA farmers systematically kill beneficial microbes with constant cultivation, fertilizers and poisons. Abundance farmers view killing microbes akin to a mortal sin because microcrops are basically made up of microbes – algae and their diverse symbiotic cousins.

The Del Monte summary report, (below) compared metrics for melon production in the abandoned land that used algae-biofertilizer with another productive melon field that used inorganic fertilizer a few miles away. Del Monte and several other companies have expanded use of these biological methods because they improve crop quality and quantity and simultaneously enhance the soil.

Blind taste-tests at ASU favored the algae-infused melons 17:1 over controls. The Del Monte professional farmer was most impressed with the 25% increase in melon shelf-life, since the melons had to be transported from Yuma to New York City.

Improves crop: • Germination rate, 20% • Time to maturity, 20% • Health and vitality, 30% • Yield and quality, 30% • Size, 20%	**Improves market value:** • Taste and aroma, 20% • Vitamins and minerals, 50 to 100% • Digestible nutrients, 50 to 100% • Color and texture, 20% • Shelf-life, 25%
Reduces production costs: • Tillage, 20% • Diesel fuel, 30% • Irrigation water, 20% • Inorganic fertilizer, 30 to 80% • Pesticide / herbicide, 40 to 80%	**Enhances soil:** • Porosity, looseness, 500% • Microbes, 500% • Erosion resistance, 50% • Bioavailable nutrients, 40% • Organic material, 20% / year

(center: SmartCultures)

Smartcultures restore degraded and abandoned soil

The melon project used terrestrial algae from the melon field. These cells thrive on land when moisture is available. They go dormant during dry times, then re-energize with rain or irrigation. Each cultivar evolved locally over eons and adapted to the unique characteristics of the *in-situ* microclimate. Yuma has a particularly brutal climate, with 125° F heat.

Algae delivered its high-nutralence cells which improved crop metrics at every life-stage. One of the best visual metrics was not reported but shows how the plants respond to algae nutralence. Mellon vines tend to grow laterally, flat in an IMA field. In the algae-infused field, plants sent shoots and flowers 32 centimeters (12 inches) into the air. The melon plants were so happy they appeared to be dancing in the wind.

Biodiversity

Algae displays astonishing biodiversity. A handful of dirt may contain several million algae cells and several thousand different species. A cup of pond water will contain even more algae cells and more species.

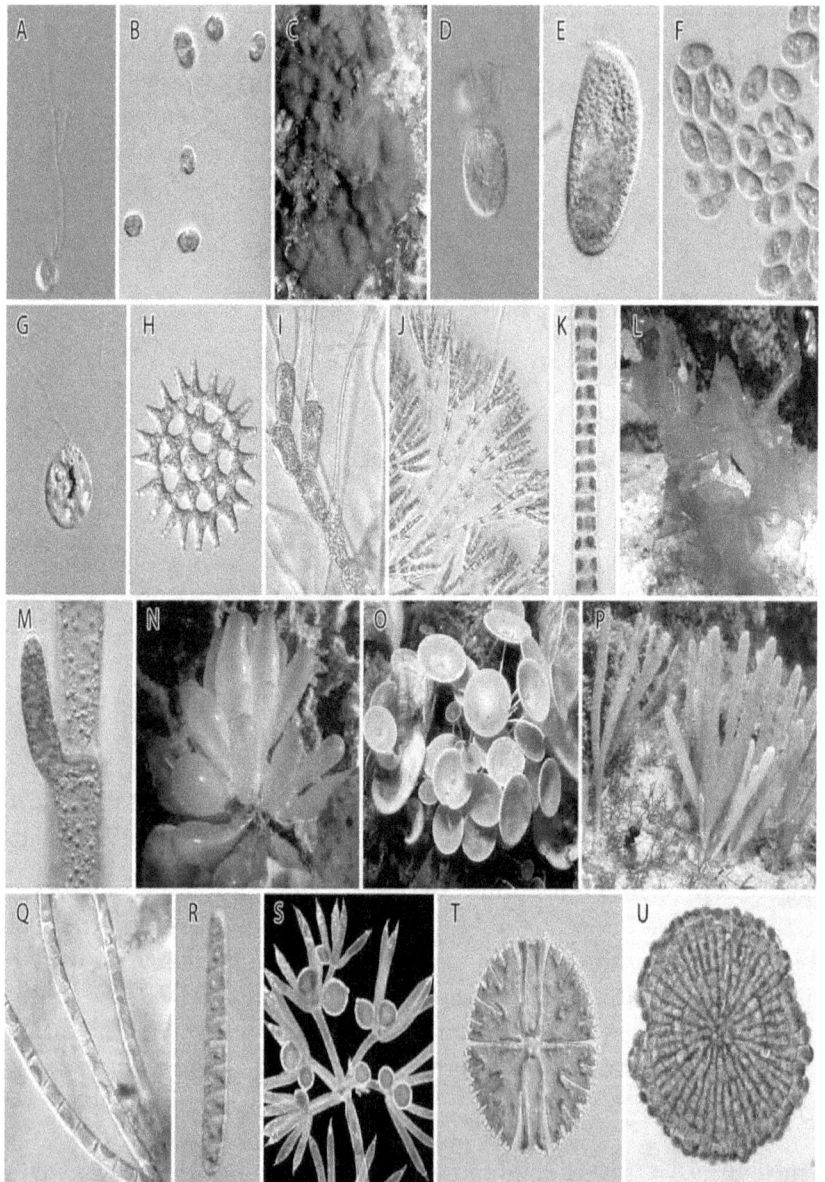

Algae biodiversity

Biodiversity enhancement occurs from multiple sources: restoration of healthy ecosystems, soil repair, clean air and water. Growers have access to the use of diverse cultivars from the 10 million algae species available.

Most species can adapt to various microclimates, but they typically thrive in the area where their ancestors have lived for millennia. Algae grow in practically infinite shapes, sizes and colors and are very adaptable.

Algae's tiny cell size create the ideal nutrient delivery system. Each tiny algae cell packages the essential nutrients to support multi-cellular life – plants, animals and humans. The algae nutrient packets are so small the algae nutrients are immediately bioavailable.

Algae, often called microscopic phytoplankton, grow in most bodies of water, moist places, on and in trees and even in rocks.[509] This tiny plant provides the foundation for the marine food chain, feeding both microbial and animal plankton; zooplankton. Subtract algae and phytoplankton from the water column and fish, shellfish, reptiles and many other aquatic creatures could not survive.

Eco-Restoration for soil

Smartcultures focuses on improving plant health and yields. Nrich™ integrates learning from Del Monte field trials and other algae biofertilizer application centered on restoring soil fertility. Nrich leverages indigenous terrestrial algae to restore soil health.

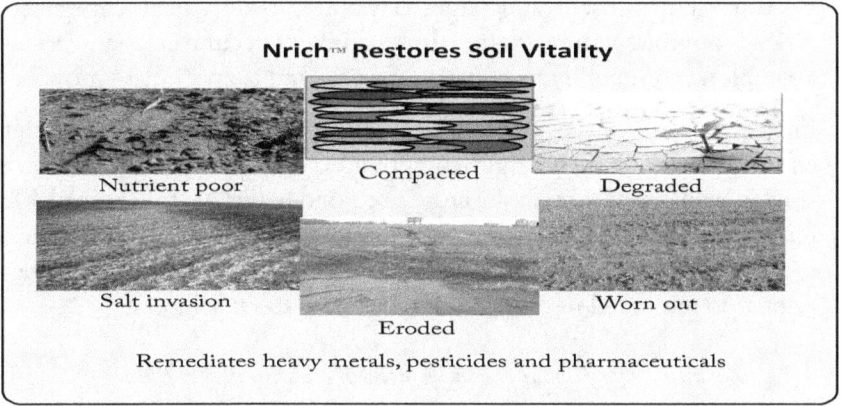

199

Nrich strategy

The core Nrich strategy cultures local indigenous algae for use as a high-nutralence delivery system for soils and crops. The algae biofertilizer made of biodiverse algae cells may be delivered through irrigation or spayed on the soil or crop. Algae cells are so tiny, foliar sprays work effectively as plants absorb nutrients efficiently above or below ground.

> *Nrich does for soils and crops what freedom foods*
> *do for people with micronutrient deficiencies.*

Cellular metabolism works the same whether growing and sustaining health in a person, animal or plant. Nrich delivers the high-nutralence that restores health and vitality to the crop's foundation – soil.

Crops respond magnificently to healthy micronutrients. Seeds germinate earlier, grow faster to maturity, make more flowers, produce higher yields and tastier produce. Algae biofertilizers increased melon nutralence significantly, which was obvious with superior size, color, nutritional density and taste.

Most freedom food cultivation strategies share two tactics that are not important with Nrich projects. Growers often cultivate a single cultivar for consistent nutrient and target compound quality, e.g. Omega-3s. Nrich biomimics nature. Nature uses a diversity of cultivars that work as a symbiotic community to provide nutrition and support plant growth.

Freedom food growers typically try to maximize productivity of their entire biomass or their target compound. Nrich growers do not have similar constraints because crops respond quickly to bioavailable nutrients. Nrich does not contribute to hidden hunger because algae are too smart – they refuse to grow if any essential nutrient is lacking. Leonardo reminded us of how little we know about soil.

> *We know more about the movement of celestial bodies than*
> *about the soil underfoot.*   *– Leonardo da Vinci*

How Nrich works

Nrich tactics adapt to local constraints, especially soil condition.

Nutrient poor soils may be repaired by algae's ability to provide the full complement of essential nutrients by special delivery. Special delivery reflects that plant immediately absorb the tiny cellular nutrients package.

Land plants cannot fix nitrogen from the air. Legumes claim that talent, but closer inspection reveals that the modules on the roots of alfalfa, soy beans, lentils and peanuts that fix N_2 are filled with cyanobacteria. The blue-green algae, cyanobacteria have the ability to fix both CO_2 and N_2, from the air or water.[510] Fixation of N_2 comes at a high metabolic energy cost, but cyanobacteria are phototrophic organisms that use sunlight to fulfill their energy demand.

Nitrogen fixation bacteria in the soil produce nitrogenase enzymes to pull N_2 from the air and transform it to plant bioavailable NH_3.[511] The ammonia is subsequently converted to nitrate by other bacteria in the soil so that it can be used by plants.

Algae biofertilizer offers a sustainable approach to synthesizing ammonia by allowing the crops to do the work. Algae biofertilizer will revolutionize agricultural fertilizer production providing a sustainable solution that cycles nutrients and reduces production costs, waste and pollution. Farmers will benefit from algae biofertilizer with higher crop yields and better-quality produce.

Compacted soil occurs from years of crushing experience under heavy equipment – cultivation, spraying, weeding and harvesting. Farmers often take monstrous heavy equipment across fields 7 to 9 times each crop production cycle. In Yuma, the red desert soil is made up of sand, hard caliche clay and small rocks. One pass with heavy equipment after a thunderstorm or irrigation compacts the desert soil severely. Equipment leaves deep soil gashes, which crushes the underlying soil structure.

Nrich expands compacted soil with two tactics. First, algae delivered as biofertilizer continue to grow in the soil as long as soil moisture remains, expanding humus and organic material. Second, some algae cells grow a polysaccharide sheath that opens similar to a ladybug's wings (but with less elegance) and creates air pockets that expand soil looseness.[512] The extra space allows trillions of other microorganisms to enter the soil.

Algae are not the singular fertility heroes. Algae produce soil honey, the plant sugar glucose ($C_6H_{12}O_6$) that attracts … just about every microorganism that likes sugar, which is every one that needs energy. The entire microbial community works together like an orchestra to support crop health.

Degraded soil occurs from years of systematic extraction by repeated crop production but with little or no replacement.

Farmers should be stewards of their land, but a recent USDA report shows that over 56% of US cropland is leased. Even worse, 80% of the lessees are non-farming landlords, who show little concern for sustainable practices.[513]

Monocrop farming degrades soil and destroys fertility by extracting humus, the vital organic matter, with every crop. Humus affects the bulk density of soil and contributes to its retention of moisture and nutrients.

The only way to replace humus published in the agri-literature, composting, is far too expensive in terms of time, fossil energy and cost. Farmers might want to use animal manure for compost, but manure is not available to two reasons. Animals and crops are raised far apart in the US, which makes transportation too costly. Animal manure contains too many pharmaceuticals, which can be transferred to field crops.

The Del Monte melon project proved that algae biofertilizer replaces humus in degraded soil. Melons are grown on a 6-foot-wide berm covered in black plastic to deter weeds and give roots sun protection.

At the end of the season, the black plastic was lifted off the berm. The red soil had transformed to green. The algae that was not absorbed by the plants continued to grow in the soil, creating rich organic material. Due to the humus accumulation, each successive melon crop, Fall and Spring, increased in yield, size and nutralence.

Nrich provides immediate relief from hidden hunger and nutrient deficiencies in soils by delivering the full set of micronutrients. Extreme soil degradation may take several growing seasons to remediate but the combination of the two tactics above can restore fertility to dead and abandoned soil. Algae's incredible ability to restore life to dead soil is only one of algae's many miracles, which are listed in Appendix II.

Salt invasion occurs when years of irrigation and fertilizer salts build up and make the soil infertile. Salt, both the lack of salt and salt invasion that destroyed soils, has spurred many wars throughout history. The words, "war" and "peace" originate from the words for salt and bread in ancient Hebrew and Arabic.

Irrigation allowed substantial expansion of crop production. A critical irrigation occurs because irrigation water accumulates dissolved salt from soil. Today, the mighty Colorado river collects so much salt that the water is too saline for irrigation when river water reaches Mexico.

Elizabeth Kolbert, in *Field Notes from a Catastrophe*, chronicled a list of sophisticated cultures that sustained themselves for hundreds of years and then crashed when their soils no longer supplied sufficient food because they had been destroyed by irrigation salt invasion.

- Tiwanaku, Lake Titicaca in the Andes – crash: A.D. 1100, drought
- Classic Mayan civilization – crash: A.D. 800, drought
- Old Kingdom of Egypt – crash: 2200 B.C., drought
- Akkadian empire – crash: 2200 B.C., drought[514]

Droughts amplify the need for irrigation. Irrigation systems designed for normal years are unlikely to deliver sufficient water in hot years when crops may need two or three times more water. Heat also intensifies soil salts, right.

Irrigation water evaporation leaves a white crust of irrigation salts that diminish and eventually destroy soil fertility.

Much of the world's food is produced in the rich soil of coasts and river deltas, below. These vital croplands are threatened by expanding cities that sprout high rises instead of food. These fertile areas are also vulnerable to rising sea levels, storm surges and seawater invasion.

River	Location	Square km	Sq miles
Ganges	Bangladesh/India	105,645	40,790
Mekong	Southeast Asia	93,781	36,209
Lena	Russia	43,563	16,820
Huang He	China	36,272	14,005
Mississippi	United States	33,670	13,000
Indus	Pakistan	29,524	11,400
Volga	Russia	27,224	10,511
Orinoco Amacuro	South America	22,500	8,700
Niger	Western Africa	19,135	7,388
Tigris-Euphrates	Southwest Asia	18,497	7,142

Fertile deltas under threat from seawater invasion

The Arctic is warming at twice the average rate of the rest of the planet which is causing polar ice sheets to melt at least twice as fast as expected. At current rates, melts are on track to raise sea levels 10 feet this century. Greenland ice is melting four times faster than 15 years ago. Antarctica is losing six times more ice mass annually now than 40 years ago.[515] The amount of sea ice loss — and the rate is acceleration — is staggering. Between 1979 and 1990, Antarctica shed an average of 40 gigatons of ice mass annually. (A gigaton is 1 billion tons.). From 2009 to 2017, about 252 gigatons per year were lost.[516] Storm surges will flood delta cropland, leaving it too saline to grow crops.

The published soil science literature offers only two ways to remove salt once salt ions have crystalized: extensive irrigation and long, but slow farmers' rain. Extensive irrigation does not work because systems do not have enough freshwater and the water already contains too much salt. Farmers' rain would dissolve the salt, but climate chaos and droughts make long, slow rain unlikely. Climate chaos brings more fierce storms that do little to remove salts dissolved in soils because the rain runs off quickly, often flooding. Most fresh water from storms is lost to runoff.

The Nrich solution works naturally to improve soil structure. Soil salt was a problem in the Del Monte project as the soil pH of 9.2 and high salt content substantially diminished soil fertility. Algae biofertilizer improved soil porosity 500%. The organic material from algae growth improved soil structure and moisture retention. Summer thunderstorms brought rain squalls, which are short bursts of intense rain. Improved soil structure allowed the rainwater to percolate deeply and carry salt ions below the root zone. After the first year of biofertilizer application, the soil recovered to pH 7.3, indicating much healthier soil.

Eroded soil occurs from years of IMA production methods as wind, rain and irrigation carry topsoil away. Farming operations that increase erosion and dust emissions include plowing, cultivating, leveling, planting, weeding, seeding, fertilizing, mowing, cutting, baling, spreading compost or herbicides and burning fields. Mechanical monocrop methods gash and crush soil with monstrous equipment.

Soil structure destruction accelerates erosion losses. Over half of the earth's farmland topsoil has already been lost, which fills and pollutes ecosystems.[517] Nature requires about 500 years to replace 25 millimeters (1 inch) of topsoil lost to erosion.[518]

At current "normal" erosion rates of 6 tons per acre, minimal soil depth for agricultural production, 150 millimeters (6 inches) is diminishing quickly. IMA methods, especially monocrops, make fertile soil a nonrenewable, endangered ecosystem. Erosion is having an economic impact because Midwest cropland recently began reporting diminished land values for farms that have experience erosion.[519]

The severe erosive force of wind and water

The U.S. net cropland losses from 1982 to 1992 covered an area the size of New Jersey.[520] About **5 million acres** (2 million hectares) of fertile agricultural lands are lost to production every year.

Neither the agri-literature nor the internet offer any practical methods for replacing a single acre of eroded topsoil. A topsoil project would first have to pay $25,000-50,000 for a civil engineering firm to do section 401 (Clean Water) and 404 (Discharge of Fill) report. Angie's List suggests replacing topsoil for a small yard at > $10,000, which makes no economic sense for cropland acres.

Two earth moving companies estimated $45,000-65,000 per acre, depending on a dozen factors. The most critical factor probably negates the entire proposition. Where would the contractor find a source for sufficient good topsoil and how much would it cost? Most fertile cropland has been farmed for years. Purposely denuded farmland would engender a value approaching zero.

Replacing cropland topsoil makes no economic sense today since good Midwest land can be purchased for $5,000 to $7,000 per acre.

Midwestern farms appear to be in a race to see whether
they run out of topsoil or Ogallala aquifer water first

Nrich does not replace topsoil. That would be a no-go mechanical solution.

A three-step biosolution takes a few years but affords fertility without moving tons of dirt.

- Install a drip irrigation system and a SmartCultures biosystem.[521] A SmartCultures biosystem is a microfarm that cultures indigenous terrestrial blue-green (to capture N^2 from the air) and green algae (to deliver other micronutrients) for use as biofertilizer.
- Irrigate the field with algae biofertilizer to improve soil structure and to attract trillions of microorganisms that work in harmony to restore humus, nutrients and fertility.
- Allow natural weeds to grow or plant a cover crop such as buckwheat, which is disked under to add to the organic material.

Soils tests available from the ag-extension service will indicate when the land has recovered sufficient fertility to support crops.

The **worn-out soil** fix is much easier than eroded soil because usually topsoil exists, but lacks organics, vitamins, minerals and micronutrients.

Monocrop farming leaves soils exhausted, depleted and sick from poisoning by synthetic chemicals and other agri-poisons. IMA methods systemically degrade soil's fertility and annihilate soils' magnificent supporting cast – microbes.

The Nrich solution uses algae biofertilizer to restore the full set of micronutrients and the organic matter that supports nutrient biosorption by the crop. Algae not absorbed as fertilizer continues to grow in the field, which improves soil structure, adds organic matter and attracts the natural microbial community.

Naturally, this Nrich algae biofertilizer solution also restores microbe biodiversity. Improved soil fertility invites biodiversity expansion as both flora and fauna feast on soil microbes.

Summary

Farmers, ag economists and politicians agree with FDR that soil destruction destroys our country. However, in spite of considerable talk, IMA practice continues to erode of cropland >20 times faster than nature can replace fertility. Unless IMA farmers make extreme changes, many will run out of fertility or water in one or two generations.

Further, surprisingly few solutions are available for restoring soils degraded or destroyed by IMA methods. While this seems like such an obvious solution, mechanical methods are simply too expensive, use too much energy and are too labor intensive. Plus, mechanical methods have not been found to work.

Smartcultures and Nrich offer biological solutions for increasing field crop yields while upgrading soil structure and fertility. While the universe of these applications is currently small, field trails have been extremely positive. Now that these biological solutions have been successfully demonstrated, smart growers can begin to diffuse these innovations.

The next chapter explores the opportunity to display graphically key differences between fossil and freedom foods.

16. Healthier for People, Producers and our Planet

Freedom foods transform the we think and talk about food and its impacts on the health of people, producers, natural resources and our planet.

Key takeaway

- How might labels convey fossil and freedom food differences?

Imagine that 10 years from now, you visit your favorite morning bar NutraU. NutraU offers a dazzling array of morning power drinks, coffee, tea and nutritionals. You make polite small talk to *N-Dow*, the NutraBot, and indicate the nutrients you want. You might name a diet or show *N-Dow* your smart phone app, iU. The iU app suggests the nutrient profile the intelligent system recommends uniquely for you.

iU remembers the form, taste and texture you like. *N-Dow* might lift some fresh fruit into your smoothie. She will add some algae components for healthy protein, micronutrients, vitamins, enzymes and bioactive compounds.

N-Dow creates your favorite morning drink that delivers all the essential micronutrients you need for the day. *N-Dow* packs your personalized drink with superb nutrition which has only a few calories sourced from healthy omega-3 oils.

N-Dow's friend, the 3D food printer bot, *N-Joy*, may make you a to-go basket with an algae-based shrimp ceviche for lunch and a healthy and tasty algae umami algae-burger for supper. Alternatively, *N-Genius* can make you a healthy surprise meal that fits your pallet and your unique health profile.

Personalized foods

Personalized foods, made to order to match your DNA, will be very attractive. Algae ingredients will make these foods easy to prepare and avoid the packaging and waste associated with fresh produce.

Smart nano-sensors will help consumers monitor their biomarkers and gut microbiome.[522] A major factor holding back consumer's access to algae-based foods is the cost of medical research. In order to prove that an algae bioactive compound actually does what medical research has proven it can do in lab studies, the FDA requires studies in three phases.

Human trials are likely to cost $50 million or more. Forbes Magazine analyzed the cost of bringing a new drug to market.[523] Their conclusion was a minimum of $1.3 billion. The average drug developed by a major pharmaceutical company costs $4 to $11 billion. No algae company has the resources for these kinds of studies.

Smart nano-sensors can cut animal and human medical trial costs by 10 to 100 times. Scientists are developing nano-sensors that enter the body, similar to an alga cell, and are transported normally throughout the body. The sensors are designed to detect biomarkers, such as enzymes, antibodies or specific cellular compounds. The sensors do their job and then pass out of the body in the urine, where they are assessed regularly by robotic urinalyses. Early sensors will detect a single biomarker, while later tools will monitor several.

Similarly, continuous non-invasive testing of body compounds from saliva, sweat or hair composition will allow fast, low-cost medical tests. Subjects might wear an appliance like a FitBit that constantly monitors several body functions, including an array of non-invasive blood tests.[524] Smart biosensors create a flood of data that needs to be filed, tracked and analyzed. Technologists will rely on Big Data, machine learning, the IofT, (Internet of Things) and AI.[525]

Implanted chips have already arrived. The first in the US, Three Square Market in Wisconsin, offered voluntary implanted chips to all of their employees in August 2017.[526]

The RFID chip, from BioHax International in Sweden, allows employees to make purchases in the micro-market, open doors, login to computers, use the copy machine, etc. Chip sensors will monitor body functions to support optimum health and medical conditions.

The FDA could play a pivotal role in algae food adoption by actively conducting human trials on 30 promising algae bioactive compounds.[527] These trials would create the metrics to assess their value in functional and medical foods. Then food companies could use their creativity to deliver excellent foods. The FDA could continue its role in monitoring producer quality. The Global Organization for EPA and DHA Omega-3s, (GOED), has done an excellent job of organizing and reporting the medical research on Omega-3 fatty acids.[528]

Algae-based meats

Algae-based meats that taste like the real thing have arrived. The most popular seafood in the US, shrimp, is made by New Wave Foods in San Francisco out of seaweed.[529] These shrimp deliver equal health benefits and taste without pollution or waste. Wild shrimp are what they eat. In their natural environment their nutrition comes from algae.

Soon, chefs will be using 3D printers to construct algae-based meats, tasting like beef, lamb, pork, goat, chicken, turkey, fish and new taste choices. When consumers can select healthier and tastier steaks that do not sacrifice animals, they will choose plant-based meat products. Microcrop meat will cost half as much as legacy animal flesh meats.

Consumers will have additional choices. Some people may select tissue cultured meats, but those products will have as many health issues as animal meat consumption.[530] Others may choose insect protein foods, but those companies must overcome a non-trivial gag factor.[531] The intriguing irony may be that both tissue culture and insect meat producers will need nutrients to grow their cultures.

Healthier for People, Producers and our Planet

Both will probably use the most popular food source on the planet that also delivers the highest nutralence – algae.

Extended food labels

Freedom foods may offer superior nutrition and taste to consumers, but how will consumers know? The Emerald Renaissance will be paired with a renaissance in food labelling the integrates valuable content and art.

Today food producers put scant information on package labels useful to health. Food producers comply with minimal FDA label requirements. Freedom foods will revolutionize food labels as food marketers will actively invent incredibly useful and artistic labels readable on smart phones, smart glasses, and personal assistants.

The extended food labels will provide the key metrics for nutralence and other valuable nutritional information. One might say: "Eating one portion of this food three times a week gives the consumer a 90% chance of avoiding heart disease, based on nutrition." Another might say: "The bioactive compounds in this food can reduce the common cold by three days or disrupt an asthma attack in one day without other medicines."

 Imagine a future where your smartphone has an app that provides the nutrient profile for foods. Freedom food marketers will create attractive QR codes consumers can scan and link to a website with information about important the food attributes.

The smartphone app will flip consumer choices quickly away from hidden hunger and empty calories to foods with high nutralence.

The website may describe how the food is different from fossil foods. A site may emphasize healthy nutrition for people, functional ingredients that support health, positive eco-footprint or other relevant benefits.

Healthier for People, Producers and our Planet

Freedom food differentiation

New foods marketers employ two primary strategies to convey their value proposition; market differentiation and market segmentation. Freedom foods naturally segment in vertical markets such as health, functional and energy foods. Freedom foods are versatile, which allows a variety of differentiation strategies. The 4P model below maximizes differences across healthier actions for people, producers, preservation of natural resources and our planet.

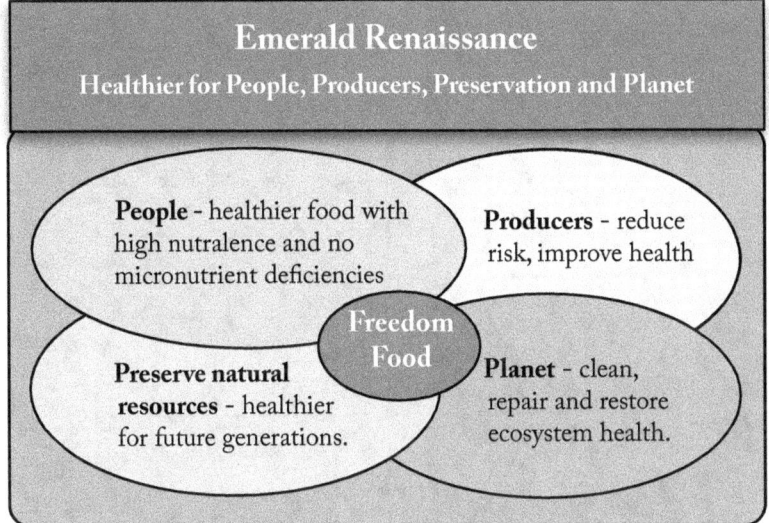

Human health is a multivariate set and can be addressed in many ways. The approach here, represents only one approach – diet. People tend to be an imperfect function of what they eat because diet does not account for the considerable variance in health from genetics, environment and exercise. Multiple, independent scientific sources have concluded that the Western diet destroys good health and the contrast between fossil and freedom foods gets the focus here.

The healthy people nutralence profile in Figure 15.1 differentiates fossil from freedom foods using health metrics that are vital to personal health. Nutrient quality displays an 8 score for freedom foods but only a 1 for fossil foods because they suffer from hidden hunger + nutrient dilution + pesticide residuals. Freedom foods are not perfect. Some lack a few useful amino acids in their protein.

Freedom food benefits for people

This is the first published metrics display showing the differences between fossil and freedom foods. These dimensions, scores and reports require further work before they will be consumer-credible.

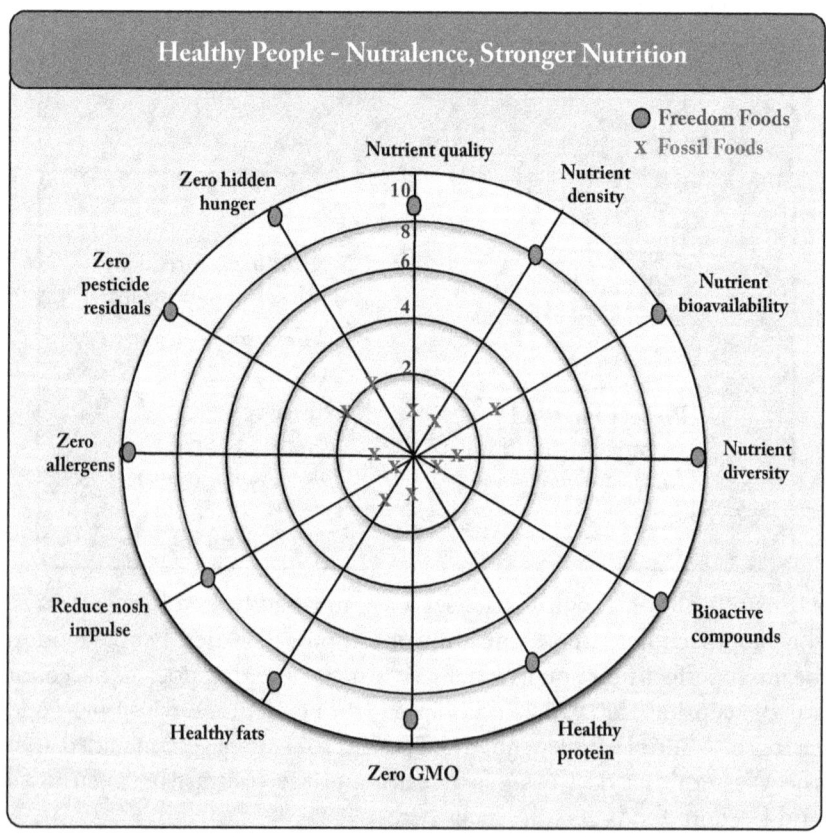

Figure 15.1. Freedom versus fossil foods nutrition comparison

All freedom foods do not need to be assessed on the same dimensions. Several dimensions are likely to change. For example, zero pesticide residuals may broaden to include antibiotics and other pharmaceuticals because they are also common food contaminates. Fossil foods are severely handicapped because no one has even theorized a way to produce crops with no contaminate dust, allergens or poison residuals.

Consumers vary widely on which attributes they value. A father of a child showing signs of obesity may be attracted to foods that reduce brain signaling and quash the nosh. Zero allergens will be appealing to a parent with a child (or family member) with allergies.

The nutralence attributes – nutrient quality, density, bioavailability, diversity and bioactive compounds – will be become more, important with additional nutrition and medical research. Science advances will share new discoveries and, equally important, refine the substantial knowledge on nutralence dimensions that benefit human health.

Nutrient quality is problematic in terrestrial crops because the nutrient value of each crop depends on the elements in the soil that are bioavailable to the roots in the current growing season. Many years of constant production strip the cropland of organic material, humus and micronutrients. Crops are likely to suffer from nutrient dilution and hidden hunger due a lack of micronutrients, vitamins, minerals and trace elements in the soil. Freedom foods overcome the nutrient quality problem naturally since microcrops grow to the limit of their available nutrients and then take a rest. Microfarmers also avoid the many chemical contaminants commonly used in producing IMA.

Modern fossil foods often deliver negligible nutrient diversity because the crops have been bred for yield (weight), not nutrient diversity. The nutrient diversity of algae foods, including over 100+ micronutrients, vitamins, trace elements and 200+ enzymes, work together to improve consumer health and vitality. Fossil foods are practically devoid of measurable levels of bioactive compounds because the crops have been bred for thousands of years for maximum yield in terms of weight, rather than nutrients of bioactives that support health.[532]

Healthier for People, Producers and our Planet

Preserving resources

Our Generation has the opportunity to leave an appalling or a positive legacy to our children and their children. This tale of two legacies was video-taped in 2019, and is meant to be played again in 2039.

> 1. *I am sorry my progeny. Our greed left you without vital natural resources. You and your family need food. I acknowledge my complicity in failing you. We consumed your life-sustaining resources.*

"The resources on which fossil foods are dependent have gone extinct. We consumed those resources to profit our family. The agri-monopoly, Big Ag, dominates how and what we plant. We did not have a choice because Big Ag dictated our production methods and our crop. We knew that staple monocultures would deplete and waste the precious resources that are rightfully yours, yet we continued our over-consumption."

"We also decided to burn food for fuel, which drove up the cost of food and left millions of own children malnourished. Burning food cut the food supply and led to the predicted food riots, followed by food wars.[533]

We knew that producing corn as a biofuel was a losers' game from the beginning. At best, corn ethanol could reduce fuel imports by less than 1%, while using more energy than it provided. Still, we extinguished trillions of metric tons of your natural resources following our folly. Our politicized energy policy wasted resources and massively polluted your world for political gain. We are slow learners. We continued burning food for fuel in spite of dozens of scientific articles showing corn ethanol to be a sham. In 2019, the US President sold out to the same sad, and mistaken story in order to hold votes from his base, the farm block."[534]

"You may ask why? In America, money buys political access, persuasion and ag policy. You can follow the money paid to Congress to understand why Congress voted to support a policy to burn food for fuel when it made no economic, energetic or ecological sense.[535] I know it seems non-sensible now, but in our time, political power trumped economics, environment and science.'

Healthier for People, Producers and our Planet

The alternative legacy provides a substantially superior story.

2. I am delighted my progeny. *Please enjoy the abundant natural resources we left for you and your restored verdant ecosystems.*

"We cannot take credit because we invented little. Nature, our teacher, showed us the way to give you the gift of sustainable and healthy food without extraction, consumption, waste and pollution.

Your grandparents and parents benefited from freedom foods that improved their health, sperm count, life quality and longevity.[536] Freedom foods gave us the liberty to cultivate wonderful foods without extracting, consuming and wasting the natural resources you deserve to enjoy. As you look around, you see verdant ecosystems that abundance methods have repaired and restored."

I only have one life to live and one legacy to give. I am thankful nature taught us the pure joy of abundance methods to cultivate freedom foods that preserve natural resources for you.

Preserving resources food label

Few consumers are aware of how incredibly fast fossil foods are extracting and over-consuming natural resources. As a consequence, few consumers factor natural resource consumption into their buyer behavior. Food labels can inform consumers and display which foods consume and which preserve natural resources.

The natural resource food label below reflects extraordinary differences in consumption. Fossil foods earn their name with a solid set of failing 1/10 ratings. Actually, a "1" rating is generous for fossil foods. The deltas, (differences) in these ratings would be even higher if the scale allowed negative ratings. A beef burger provides a good example.

The USGS estimates that production of a single hamburger consumes 18,000 gallons of water.[537] The European Commission's "Generation Awake!" campaign created a neat video, using water balloons to illustrate the exact amount.[538] In contrast, a freedom food plant-based burger adds 20% more fresh water as a coproduct of microcrop production.

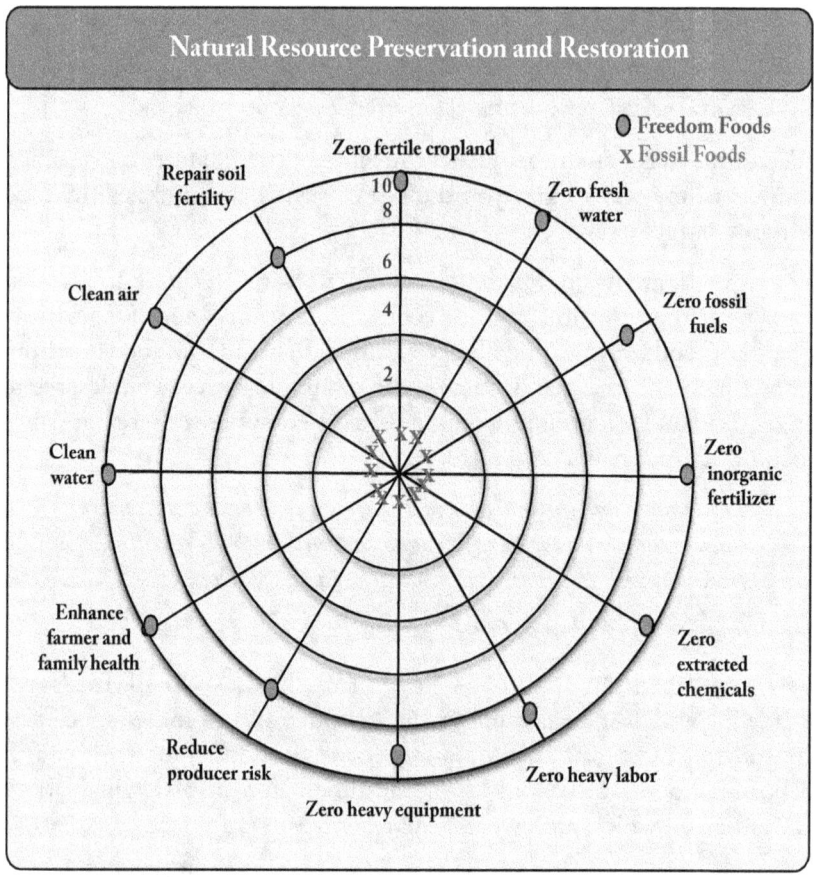

Benefits for our planet

The healthy planet label aligns with the Green New Deal.[539] This label can be modified to align precisely with the Green New Deal goals. The label reflects avoidance of environmental waste and pollution and the restoration of air, water, ecosystems and biodiversity.

The healthy planet label provides a visual display of fossil foods' linear model that uses natural resources only once, inefficiently, and then discards them into our environment. Freedom foods reflect a circular economy. Resources are biocycled, creating zero waste and pollution.

Healthier for People, Producers and our Planet

Each dimension represents substantial value. Zero air pollution means that freedom food production zeros-out the massive CO_2, methane, nitric oxides, black soot particulates and other GHGs emitted producing fossil foods. Freedom foods not only avoid GHG pollution but capture millions of tons by recycling those elements into healthy bioproducts.

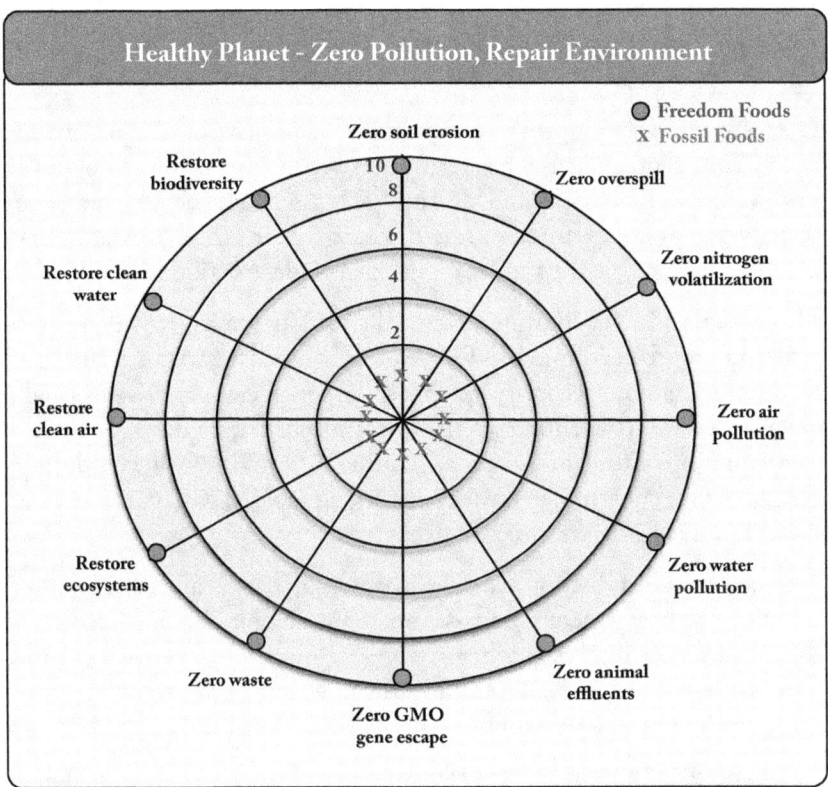

The accrued value of GHG avoidance, capture and biocycle to human health – asthma, allergies and airway diseases – will save tens of billions of dollars a year. The process will also mitigate global climate chaos.

Imagine the value of a food production system that consumes zero fossil fuels while also removing fossil fuel pollution from our atmosphere. The accrued value to human health alone would be worth >$600 billion a year.

The value of avoiding clean water consumption and remediating polluted water may have more value in terms of total savings than air remediation. IMA contributes about 80% of the water pollution in the US, creating over $100 billion a year in externality costs. Freedom foods can save those costs two ways. Microfarms can mitigate at least half of IMA overspill and pollution. Microfarms can also cultivate healthy food without any freshwater and create 10% extra fresh water in the multiproduct process.

Clean water globally provides even more value. Every year more than 3.4 million people die as a result of water related diseases, making it the leading cause of disease and death.[540] *The Lancet* concluded that poor water sanitation and a lack of safe drinking water take a greater human toll than war, terrorism and weapons of mass destruction combined.[541]

Heavy metals in groundwater, especially lead and arsenic, will generate 100 times more headlines in a few years as aging infrastructure fails, lead in water pipes, and wells are extended deeper. Heavy metals, especially arsenic, tend to concentrate in the lower reaches of aquifers. Arsenic is particularly plentiful in American aquifers. Drinking water and crops irrigated with contaminated water and food prepared with contaminated water are the primary sources of exposure.

Microcrop bioproduction can clean water of heavy metals, including arsenic. The heavy metals biomass can go into bioplastics, bio-resins and bioconstruction materials where they stay inert and non-threatening. The same biosystem can produce fresh water as 10 to 30% of its multiproduct production mix.

Meat and dairy products score poorly on eco-food labels. Each gallon of dairy milk consumes 880 gallons of water. California's 1.8 million dairy cow produce over three billion gallons of milk and 35 million tons of manure.[542] Manure emits extensive amounts of nitric oxides and ammonia. Cows burp and fart daily 6,137 liters of CO_2, 5,279 liters of CH_4 (methane) and lots of ammonia, depending on their diet. Imagine that each cow's gas daily would fill 11,400 one-liter soda bottles.

Healthier for People, Producers and our Planet

Fossil food staples, corn and soybeans scores are uniformly 1/10. They flunk on all the healthy planet dimensions. When freedom food providers share their eco-labels, consumers will not only avoid fossil foods but demand that fossil farmers pay for their waste and pollution.

Food labels that score the degree to which fossil foods contribute to healthier people, producers, natural resource preservation and our planet show that fossil foods flunk across the board. These appalling scores refresh the question poised earlier:

> *Why should taxpayers support the Big Ag farm policy that fails to deliver healthy food, puts farmers at severe risk, over-consumes natural resources, and pollutes our environment?*

The answer for decades has been: "Big Ag continues because another food production model has not come forward to challenge fossil foods." Those days are past. Freedom foods are superior to fossil foods an all human and producer health, economic and sustainability dimensions.

Additional differentiators

The differences between fossil and freedom foods repeat the 8 to 9-point deltas with additional assessment dimension sets. Scores on **producer health** below, reflect the substantial physical, health and economic risks fossil foods impose on farmers.

Freedom Food Label Dimensions - Producers	
Producer health	**Farm family health**
• Physical risk - equipment	• Clean air
• Health risk - exposure	• Clean water
• Economic risk	• Zero agri-poisons
• Heavy labor	• Clean community
• Fatigue	• Farm animal health

Are those risks acceptable? No, they are not. But they are unavoidable with IMA methods. Fossil foods cannot be produced without substantial risk to the producer, farm animals and farm communities.

Abundance methods mitigate farmer, farm family, and rural community risk. Freedom foods offer new opportunities for diverse microfarm operators in almost any geography. Microfarm engagement offers equality in cultivation for diverse people. those with developmentally disabilities, disabled veterans, the elderly and others.

Bioactive compounds

Bioactive compounds are probably the most significant differentiator separating fossil and freedom foods. Terrestrial crops must distribute their limited energy across so many physical features, little or no surplus energy remains to produce bioactive compounds. Bioactives exist in microcrops because single-cellular organisms have had to fight for life solo. Over billions of years, microorganisms wisely developed a superb cache of defensive and offensive weapons called bioactive compounds.

Why bioactives?

Each alga cell had to fight alone for its survival and opportunity to propagate over eons. Cells could not rely on superior size or strength like multicellular plants. Consequently, bioactive compounds evolved in surviving cells that provide defense against malicious attacking organisms. Other bioactives provide mechanisms for treatment if threat vectors penetrated the cells' outer defenses.

Fossil farmers protect their crops with external-to-the-organism tactics. Farmers have safeguarded field crops for thousands of years, typically using physical/mechanical rather than biological solutions. Farmers used many solutions that required considerable human or animal labor; plows, disks, hoes and shovels. The Green Revolution reduced physical labor as farmers employed large mechanical equipment for field work and applied chemical poisons intensively for crop defense against predators.

The unintended consequence of farmers' attentiveness is that terrestrial crops have lost the need to produce bioactive compounds for their own defense. Similarly, IMA crops have few residual bioactives to leverage natural biota to increase growth speed, nutrient density or yield.

Modern crops display the plant form of "learned helplessness." Why should they expend the energy necessary to produce bioactives? They have learned that they can depend on farmers for precision seeding, weeding, cultivation, irrigation and fertilization. Fossil crops' natural defenses have atrophied from disuse. IMA farmers supply predator protection with millions of tons of herbicides, pesticides and fungicides.

Proof of how bioactive compounds have withered away in fossil crops shows in bioassays. Crop metrics indicate that most of over 200 bioactive compounds prevalent in lower-level plants like algae are either no longer in the rooted plants that evolved from algae, or the bioactives fall below detectable levels. A few bioactive compounds are listed below.

Freedom Food Label Dimensions - Bioactives	
Bioactive compounds	**Bioactive componds**
• Antioxidants	• Sulphated polysaccharides
• Soluble fibers	• Phlorotannins
• Proteins and amino acids	• Carotenoids
• Minerals and trace elements	• Peptides and sulfolipids
• Vitamins and phytochemicals	• Antivirals, antifungals
• Polyunsaturated fatty acids	• Antibacterial

Bioactive compounds will become 100 times more salient with automated detection, additional medical research and food labels.

Avoidance of legacy foods

Today, over 93 textbooks are available in English for courses in Food Marketing and Consumer Behavior. Not one textbook suggests consumers should avoid fossil foods. Of course, many cite popular diet alternatives. Texts ignore how food production affects the health of producers, natural resource preservation or our planet. The excellent work by the EAT-Lancet team should spur changes in future editions.[543]

Healthier for People, Producers and our Planet

Many consumers will adopt freedom foods in order to avoid the health problems with fossil foods that inflict people with the diseases associated with the Western Diet.[544] Consumers interested in promoting their own and their family's health and longevity will become freedom foods advocates. Microcrop foods will provide extra years of life proven by vegetarian diets, without the threat of empty calories or allergens.

Millions of people have allergic reactions to food and treatments costs US families over $25 billion a year.[545] Food allergies can cause severe reactions and may be life-threatening. The FDA recommends **strict avoidance** of food allergens.[546] They note that early recognition and management of allergic reactions can prevent serious health consequences. The FDA and the University of Nebraska maintain sites with details regarding the occurrence and severity of food allergies.[547]

Any parent that has taken a child to do a series of 28 anti-allergy shots will become an enthusiastic supporter of freedom food. After each shot, the child must be monitored for 30 minutes for possible reactions. The FDA requires 14 food allergen sources to carry food label warnings.

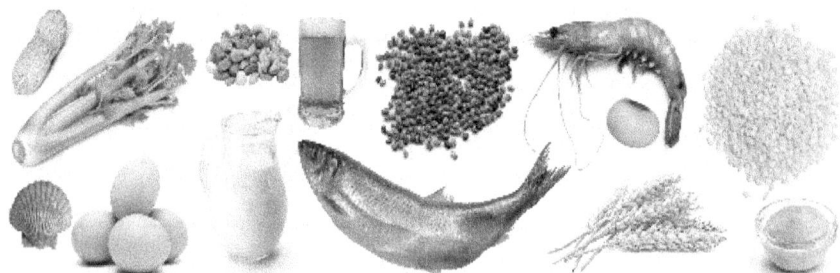

FDA research **reports** there is no cure for food allergies.[548] Algae nutrients can replace all the nutrients in the 14 allergen foods with no allergens. Algae sit on the lowest rung on the food chain and produce no known allergens.

New meat allergies have arrived, and this tick carries a very nasty bite, (right). The lone star tick cometh to change or terminate lives for thousands.[549] Red meat contains several protein-linked saccharides: galactose-alpha-1, 3-galactose, and alpha-gal.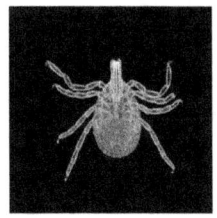

Symptoms include itching with angry hives blossoming, stomach cramping, fainting and for some, death.

Microcrop organic and vegan meats will provide a safe alternative that continue to give the familiar texture and taste sensations that make animal flesh attractive. Smart consumers will see the benefit of foods that do not cause allergic reactions, for themselves and for their children.

Counter arguments

Harmful algae blooms (HAB), can produce toxic substances, e.g. red tides.[550] Extensive research has led to a classification of which species and under which conditions these toxins are created.[551] Freedom food producers avoid those species and cultivation conditions.

One of the most popular arguments against microcrop foods is: "I want to know what I am eating." If this were true, consumers would never eat chicken nuggets, hotdogs or other processed meats.[552] Those meats are made from offal, organ meats, entrails and contain dozens of unpronounceable compounds.

Can fossil food producers change?

Several sources including the UN *Global Nutrition Report* suggest that fossil food producers should mitigate their consumption, waste and pollution.[553] The healthy planet metrics reveal two strategic elements:

1. Fossil foods producers face such a huge delta, how can they possibly overcome the multidimensional disadvantages caused by the linear IMA food production model?

They cannot. No theoretical model exists that shows how IMA methods can reduce consumptions, waste and pollution by even 10%. GMO crops increase, not decrease, across every consumption and waste dimension. Most scientific reports show key indicators – GHG, water pollution, biodiversity loss and ecosystem degradation – are systemically increasing.

2. Starting fresh with freedom foods overcomes the substantial fossil food health disadvantages for people, resources and our planet.

Abundance methods and freedom foods stop systemic consumption, waste and pollution. Abundance methods not only reverse the negative environmental trend lines but also repair and restore air, water and soils.

Summary

Food labels clearly display significant differences between fossil and freedom foods on each of the 4Ps. Mechanical fossil food production fails on every dimension while freedom foods scores soar. This summary food label clearly shows which method provides the best way to produce a healthy, clean and sustainable food supply. The metrics humble our human brains because we came up with mechanical methods while largely ignoring nature's genius.

These labels convey >20 times more important food information than the FDA requires for foods today but are unlikely to be required in our lifetime by the notoriously slow and pedantic FDA. Discussions with FDA officials resulted in the observation that the freedom foods health information seems "too specific."

The FDA excused itself from food label information on resource preservation, farmer and environmental health, because those issues fall outside the FDA jurisdiction.

New food labels will be driven by smart consumers who demand food sourcing transparency. Freedom food companies will create informative labels because the content conveys how new foods are superior to fossil foods. Several food production methods are superior to fossil foods on most the critical dimensions, including hydroponics, aquaponics, aeroponics, organic, resilience, as well as abundance.

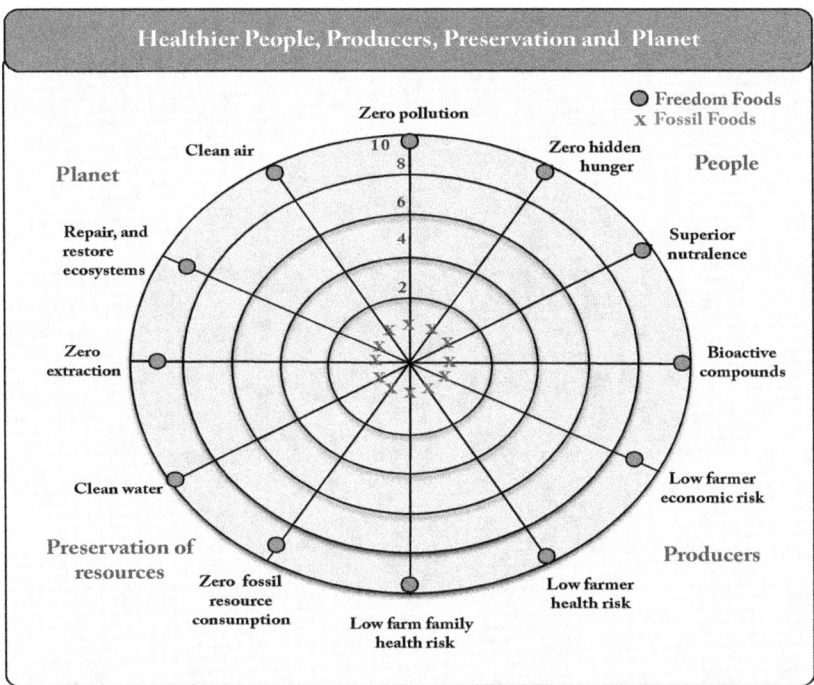

Consumers will demand these labels and only buy food products that show how they were produced and what positive or negative 4P health impacts the food provides.

Oh yes, price!

Price remains the most critical marketing variable for many consumers evaluating new food choices. Affordable prices increase demand and access by impoverished consumers. High prices are show stoppers.

> *New food production methods have a near-zero chance to*
> *flourish in America due to the USDA monopoly power.*

Hidden government subsidies artificially decrease the prices for fossil foods by at least 60%. When medical costs from fossil food health impacts are factored in the full accounting, fossil foods are artificially underpriced by 80%. The next section explores full cost accounting for our current foods supply chain.

Healthier for People, Producers and our Planet

17. What do Fossil Foods really Cost?

The hidden billions in taxpayer fossil food subsidies disguises the true cost of food and creates a deceitful clear and present danger to all consumers. Even worse, subsidies prevent clean and healthy food cultivation technologies from gaining traction.

The first question people ask when considering freedom foods is:

If freedom foods are so great, why are they more expensive?

Answer: *Freedom foods' true cost is less than 20% of fossil foods.*

The answer has been carefully hidden from the public. When accounting for hundreds of billions of dollars in hidden costs, freedom foods net cost is substantially lower, 80% less, than cost than fossils – today. This may be the most incredible story in food history.

Freedom foods appear to be more expensive than fossil foods based on current commodity prices. Appearances can be deceiving, like an iceberg. A full cost accounting shows many of the costs are occluded from public view.

The Farm Bill, Farm Bureau and Big Ag have carefully hidden seven dishonesties from the American public for decades.

While they trumpet the low cost of modern food, they are simultaneously lobbying for ways to transfer even more wealth to farmers as a means to disguise accurate fossil food prices. The hundreds of billions of dollars in direct and indirect farm subsidies must be the **largest wealth transfer in American history**. Most subsidies are funneled to a few rich white, male mega-farmers. Even if subsidies made sense, which they do not, distribution lacks any form of equity.

What do Fossil Foods really Cost?

The seven deceits create a façade that make fossil food appear to have low prices. American taxpayers fund these deceits with billions – to our own detriment. Food costs should be transparent to consumers.

Seven Deceits not Reflected in Fossil Food Prices

The true cost of food should include these hidden costs.
1. Hidden hunger inflicts catastrophic medical costs.
2. Hidden farm cash subsidies.
3. Hidden executive orders.
4. Hidden non-cash farm subsidies.
5. Hidden energy policy subsidies.
6. Hidden natural resource consumption.
7. Hidden environmental impact costs.

Each deceit falsifies the true cost of fossil foods to consumers.
Each deceit benefits fossil foods to the detriment of all new foods.
All food costs should be transparent to consumers.
Best solution: End all subsidies and let new foods compete.

Few taxpayers will be willing to fund these billions when they knew the full impact of each of these deceits. These seven deceits drive health degradation for people, producers and our planet.

Cost accounting for the full cost of fossil foods will create a firestorm. Good, let's have an open debate. My data sources include:
USDA, although their site hides this topic.
Environmental Working Group, spent years working through public records requests to acquire their superb data set.
Extensive science and agribusiness research and publishing.
Hundreds of interviews with food business, science and policy leaders during 50 years of learning, teaching and consulting in all levels of our global food supply chain.

What do Fossil Foods really Cost?

Subsidy takeaways

Subsidy discussions were fascinating and generated passions.

- Well over half of all farmers and most others in the food supply chain believe subsidies are a waste, providing more pain than gain.
- Subsidies pay Big Ag to produce more tonnage, which amplifies micronutrient deficiencies. USDA subsidies pay producers to intentionally damage human health, which is unconscionable.
- Subsidies hurt and often destroy small US farmers and farmers in Mexico, Canada and Central America. US farm policy decimates domestic and foreign small farms by artificially undercutting prices.
- Subsidies discourage US farmers from innovating, cutting costs, diversifying, and limiting their use of fossil resources. Instead, subsidies encourage gross waste and environmental abuse.
- Young people understandably detest subsidies because transfer payments eliminate any option they have to go into farming.

Congress should end all farm subsidies, possibly leaving some protection for crop insurance. Businesses in other industries face risks, yet they succeed or fail based on wise practice. They are not shielded from poor practice by insurance and federal subsidy cushions.

Seven deceits

The full cost accounting for fossil foods has not been published before. The EWG and others have done an excellent job are quantifying many of the subsidies that should be publicly transparent but are not. Many people who know the truth cannot publish because it would be a death wish for their academic career. Kate Cox in the *New Food Economy*, did an superb exposé on how the USDA and Farm Bureau effectively stifle research and hide truths that do not square with their business models.[554]

The USDA uses the Big Tobacco strategy; "Our foods do no harm." Big Ag effectively quashes scientific research that shows the negative impacts from fossil food production. They block publishing on critical topics like animal effluent damage, GMOs, monocrop destruction to cropland and biodiversity, and pesticide and poison damage to human health.[555] Kate Cox understated her exposé that discredits the honesty and values proffered by the USDA, Big Ag and the Farm Bureau.[556]

Key faculty members would not speak to her on record. They knew the predictable consequences that have torpedoed many promising careers.

The seven deceits confront the USDA's false actions directly. The USDA, et al, are welcome to publish a defense that should include a full accounting for each of the deceits exposed here. These deceits put their children and their grandchildren in extreme health jeopardy.

In marketing, when a company intentionally keeps any key product or service attributes hidden from the public, that company deceives. Typically, free markets allow legal consequences brought by other market competitors. Agribusiness markets are not free. The Big Ag monopoly rules food policy and the Farm Bill. Who dares challenge the US government?

We, the public, can challenge that the USDA intentionally deceives about product – "fossil foods are good for you," and price – "fossil foods are cheap." Fossil foods are cheap only if we ignore the massive hidden subsidies that artificially reduce food prices at the grocery store.

These deceits provide a means to examine the full cost of fossil food. The most expensive hidden cost from fossil foods occurs because the primary product of IMA – foods – are nutritionally deficient, deliver empty calories, and seriously damage human health.

Hidden hunger

Deceit 1. Hidden hunger inflicts catastrophic pain from imposing micronutrient deficiencies that lead to preterm deliveries, autism and other developmental disabilities. Hidden hunger causes obesity, diabetes, CHD and cancers. The Western fossil foods diet has severely damaged the health of American and global families.

> Unhealthy diets *now pose a greater risk to morbidity and mortality than unsafe sex, alcohol, drug and tobacco use combined.*[557]

What do Fossil Foods really Cost?

IMA knowingly cheats consumers and the public because the impacts of hidden hunger have been known in the science community for years. Hidden hunger conceals nutrient deficiencies which are transferred from field crops to American children and adults. Many have no choice but to live on the fossil foods diet. A CDC report found that 9/10 Americans live in a food desert, without access to fresh, whole foods.[558]

Deceptively low food prices do severe harm. Low prices make fossil foods appear cheap to poor consumers. Food-related medical costs now exceed 20% for many families. These medical costs are not reflected in fossil food prices. If any other industry inflicted this cost and pain on consumers, that industry would be shuttered, (except possibly Big Oil).

Total US health expenditures were $3.3 trillion ($10,348 per-capita) in 2016.[559] Assume 33% of those expenditures were fossil food related. Then, **$1.1 trillion should be added to the real cost of fossil food.**

Hidden direct cash subsidies to farms

Deceit 2. Hidden farm direct cash subsidies more than double the true cost of food.

> *Americans get hit three times: once spending dollars at the grocery store, second, spending more on health care and once more by sending billions of dollars in tax subsidies to Big Ag.*

The billions disbursed for farm subsidies are paid by the same consumers who are forced to live on fossil foods with micronutrient deficiencies. These billions should be, but are not, reflected in fossil food prices.

Between 1995 and 2010, farm subsidies averaged $52 billion a year. Some of these billions subsidized four junk food components: corn syrup, high-fructose corn syrup, corn starch, and soy oils.[560] It is difficult to comprehend why the federal government subsidizes food that directly contributes to America's epidemic of obesity and diabetes.

Subsidies are hard to justify when farmers' average income in 2016 was 42% higher than the average of all households; $117,918 to $83,143.[561] IMA is no riskier than other industries and does not need costly and hidden taxpayer subsidies.

What do Fossil Foods really Cost?

Subsidies lavishly over-reward wealthy farmers. Over 60% of subsidies from the three largest programs go to only 10% of the largest farmers, **who are all are white, male, million or billionaires.** Between 1995 and 2016, the largest 10% of farmers received 77% of subsidies. The top 1% received 26% or $1.7 million per recipient.[562]

Direct cash subsidies to fossil foods farmers include over 60 programs.

Subsidies	Purpose	Cost / year
Crop insurance	Insure against crop losses. >70% go to corn, soybeans and wheat.	>$8 billion[2]
Agriculture Risk	Assures revenue per acre does not fall below a guaranteed level.	$3.7 billion[17]
Price Loss Coverage	Assures crop reference price. Prices are set high, which makes payouts likely. Farmers can double dip from multiple subsidy programs.[19]	$3.2 billion[18]
Conservation Programs	The USDA runs numerous farm conservation programs. Several work against conservation.	>$5 billion
Marketing Loans	Guarantees prices with a loan at harvest so farmers can hold crops to sell at a higher price later.	$160 million[21]
Disaster Aid	Insures against disasters – wildfires, hurricanes, etc.	$1.9 billion[22]
Marketing and Export Promotion	Farm and food promotion, especially exports, only for fossil foods.	$1.2 billion
Research and Support	Agriculture, food R&D, statistical data and economic studies.	$3 billion

What do Fossil Foods really Cost?

Only eight are summarized above. They tell the story. Subsidized crop insurance encourages farmers to maximize subsidies, not market efficiencies and certainly not nutrition. Subsidies induce farmers to take unwise risks, since taxpayers pick up the tab whether they succeed or fail.[563] Incredibly, subsidies actually pay farmers to intentionally harm the environment and to waste our precious natural resources.

Subsidies create incentives that do extensive environmental damage. Cop insurance subsidies encourage farmers to expand crop production on highly erodible land. Wetlands, pasture, grazing and forest lands are changed to crop production, creating severe environmental harm.[564]

Subsidy fuzzy math goes this way. The USDA subsidizes crops against yield loss, which creates an incentive for farmers to use marginal land, such as wetlands, grazing and forests. These are sure to provide subsidy income from low production. Then, the USDA pays another subsidy to "conserve," which gives farmers additional incentive to use marginal land one year for weak yield subsidies and then claim conservation subsidies the next. There are no audits or penalties on cheaters and recidivists.

USDA crop subsidies destroyed Haitian, Mexican and Central American farmers because they could not grow food stables as cheap US food dumped on their country as "food aid." Subsidies have displaced over a 2 million poor Mexican farmers. Farmers were forced to leave their land because they could not compete with subsidized US food. Many of these farmers added their feet to the flow of illegal immigrants to America. Canada, Mexico and other countries have outstanding lawsuits against US subsidies with the World Trade Organization because these subsidies artificially depress the real price of food grains.

Total subsidy payments, below, for only four of over 60 programs are stunning. The key question should be:

> *Why should taxpayers subsidize millionaire agribusinesses when their actions do harm to consumers, farmers, farmlands, and our environment? Subsidies also hide the true cost of food.*

What do Fossil Foods really Cost?

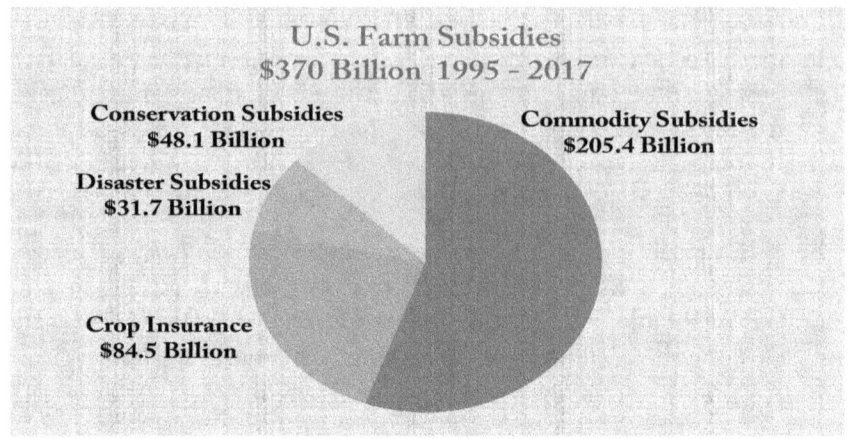

Farm Subsidies by the Numbers – Source EWG[565]

Direct cash subsidies to farmers of at least **$40 billion annually** should be added to the cost of fossil food.

Hidden executive order subsidies

Deceit 3. Hidden executive order subsidies add substantially to the true cost of specific foods.

Agriculture Secretary Sonny Perdue announced in August 2018 that the USDA will authorize $12 billion and possibly more for trade damage payments to American farmers. About $3.6 billion will go directly to soybean farmers to offset their losses from Trump's tariffs on China. In 2016, the USDA purchased 11 million pounds of surplus cheese for $20 million as part of a "robust, comprehensive farmer safety net." In 2019, dairy producers are requesting another bailout because consumer cheese tastes have changed. The USDA estimates coolers currently store 1.4 billion pounds of unsold cheese. Meat and poultry producers in 2019 are requesting a bailout to buy 2.5 billion pounds of surplus meat.[566]

What other industry receives a similar safety net? The auto industry received a one-time capital infusion, which was less than annual fossil farm subsidies. The auto industry monies were repaid. Agribusiness subsidies are recurring annually, with no repayment requirement.

What do Fossil Foods really Cost?

Hidden executive order subsidies to farmers add at least **$3 billion annually** to the cost of fossil food. This assumes, without proof, that the $12 billion plus trade damage payments in 2018 remain an anomaly of the Trump administration.

Hidden farm non-cash subsidies

Deceit 4. Hidden farm non-cash subsidies more than double the true cost of food.

The incredibly high cost of capturing, storing and moving water for irrigation has been estimated at $430 dollars per acre foot. A California farmer may pay $100 per acre foot, which is about 0.03 cents per gallon. San Francisco residents pay about 1 cent per gallon, or 33 times more than farmers. Taxpayers fund the farmers' water discount.

California's water overdraft imbalance rate equals 1.8 million-acre feet per year. The San Joaquin Valley's region's aquifers are losing nearly 600 billion gallons of water each growing season. Over the past three years, Central California farmers drilled over 5,000 wells, going thousands of feet deep to reach thousand-year-old aquifers before their neighbors did in a groundwater arms race.[567] In many areas, waiting time for deeper wells now exceeds two years. California regulators have set a goal for net zero aquifer loss by 2040. No regulators seem to want to go on record to assure that Central Valley aquifer water will be available in 2040.

Over-drafting aquifers causes a litany of problems; water contamination from agri-chemicals and heavy metals, high energy costs, and the loss of drinking water in rural communities. Before an aquifer crashes, hydrology interferes, making pumping water from deep underground too expensive. As wells go deeper, more dangerous heavy metals are likely to add to the agri-poison contamination. A March 2019 report shows the most productive US region, the San Joaquin Valley will have to leave over half a million acres fallow to sustain water for people.[568]

Irrigation water represents the largest public subsidy to agriculture. Federal, state and local governments pay for design, construction and maintenance of all of the catchments, waterways, reservoirs, lakes, dams, pumping stations, aqueducts, pipelines, tunnels, culverts, bridges and canals that deliver irrigation water. Even the ditch tenders, who turn the water on and off for farmers, are subsidized by local water utilities.

A Public Policy Institute study estimated California pays over $30 billion a year to manage water.[569] The public has no idea how much irrigation subsidies cost because those costs are not segmented in water reports. Farmers use 80% to 90% of the freshwater in Western states while contributing less than 2% to GDP. Many farmers sell their meat, dairy, nuts, fruits and food grains both in the US and abroad, taking advantage of the public gift of irrigation water. Taxpayers subsidize irrigation water from which farmers, not taxpayers reap substantial benefits. Of course, some of the cost would have to be incurred to provide sufficient city water. Most rural household water comes from well-water, which costs more as wells must go deeper.

Some estimates calculate that if farmers had to pay as much as city residents for water, the price of food would triple. Another calculus yields a different solution.

If farmers did not benefit from massively subsidized water, they would be motivated to change to hydroponics, aeroponics or abundance methods which require minimal freshwater.

Fossil fuel costs are hidden by Big Oil direct subsidies of $4.5 billion a year.[570] About $1 billion of the Big Oil subsidies provides tax exemptions for farm fuel, another substantial, but hidden farm subsidy.

Big Oil direct and indirect subsidies have been estimated to save farmers, (as well as all consumers), $5 a gallon at the pump. These fossil fuel savings are enjoyed throughout the entire food supply chain, including all the agri-inputs, seeds, cultivation, seeding, weeding, fertilizing, harvesting, storage, processing, packaging, transportation, and retail.

Saving $5 a gallon may not seem like much, but farmers use about 6 gal/acre of diesel fuel, which yields $30 savings per acre. The profit farmers made in 2018 on soybeans was $30 dollars an acre.

What do Fossil Foods really Cost?

Credible sources estimate Big Oil's indirect subsidies sum to $40 billion a year when the many tax breaks are added to the equation.[571] Here are a few examples compiled by Taxpayers for Common Sense.

- Volumetric Ethanol Excise Tax Credit $31 billion
- Intangible Drilling Costs $8.9 billion
- Oil and Gas Royalty Relief $6.9 billion
- Percentage Depletion Allowance $4.3 billion
- Refinery Equipment Deductions $2.3 billion

The hidden costs of irrigation water and fossil fuel subsidies should add at least $50 billion a year to the true cost of fossil foods.

Hidden energy policy subsidies

Deceit 5. Hidden energy policy subsidies add substantially to the true cost of a wide variety foods, especially those that that use compounds from corn or soybeans.

All farm subsidy programs use fuzzy logic and even fuzzier math to intentionally fool the public. Energy policy subsidies are the fuzziest on both logic and math. Here is the incoherent logic.

The UN Intergovernmental Panel on Climate Change calculated the CO_2 and climate benefits from replacing petroleum fuels with biofuels like ethanol are zero, (and implicitly negative).[572] The report identifies many negative ethanol impacts such as direct conflicts between food and fuels, land-use changes, water scarcity, loss of biodiversity, fertilizer scarcity and nitrogen pollution through the excessive use of fertilizers.

The argument that ethanol saves fossil fuel imports is not supported by data. A liter of corn-based ethanol contains 5,100 kilocalories of energy – but it takes 6,600 kilocalories worth of fossil fuels (diesel for the farm machinery, petroleum to make fertilizers and pesticides, etc.) to produce enough corn to make a liter of ethanol.[573] Corn-based ethanol does not save fossil fuels. Corn ethanol production wastes fossil fuels while destroying valuable cropland, water and ecosystems.[574]

What do Fossil Foods really Cost?

Rising prices of corn feed have driven many animal ranches out of business. As grocery prices increase, so does federal spending on programs, especially the Supplemental Nutrition Assistance Program. Ethanol-induced food price increases disproportionally affect the poor.

Ethanol math is hidden from the public because subsidies are tucked in a labyrinth of dispersed programs. Tax credits through 2012 cost the US Treasury over $40 billion. Over 60 other federal and state programs allot billions of dollars to ethanol in various forms; import tariffs, infrastructure subsidies, grants, loan guarantees, tax credits, and other subsidies.[575] Direct ethanol subsidies from 2010 to 2013 were $15 billion every year.[576]

Under the 2007 Independence and Security Act, Congress mandated that the US use 36 billion gallons of ethanol biofuels by 2022. The federal government both requires the use of ethanol and subsides it, so farmers do not lose money producing a loser biofuel crop.

Corn ethanol subsidies whack taxpayers six times. Taxpayers pay more:

1. At the grocery store for higher-priced food because 40% of US corn burns as fuel. The Congressional Budget Office estimates consumers spend more than $3.5 billion more on groceries annually due to the ethanol mandate.

2. At the gas pump because mandated ethanol costs more.

3. At a quick return to the gas pump because ethanol has only 66% of the energy of gasoline. The Institute for Energy Research calculated that Americans have to spend an extra $10 billion per year for fuel.

4. At a quick return to their mechanic because ethanol erodes and damages engines. Many vehicle companies now state that ethanol blends above 10% void all engine warranties.

5. With utility payments as utilities must spend $4.8 billion annually to remove corn nitrate pollution from drinking water supplies.[577]

6. At federal tax time to pay for the myriad of ethanol subsidies.

What do Fossil Foods really Cost?

History may designate corn ethanol subsidies as the worst mistake the US Congress has made in 200 years. Burning food is a vicious strategy of war, designed to punish the enemy, starve their children and destroy their will to fight. America is the first country in history to intentionally harm our own children by burning their food. Burning food reduces supply and drives up food costs for everyone. The over-consumption of non-replaceable natural resources and pollution of air, water, soils and ecosystems creates incalculable loss and irreparable damage.

The maze of corn ethanol subsidies has allowed the federal government to pick winners and losers, distort energy and agriculture markets, and expand expensive overproduction of corn, soy and ethanol.

The cruelest ethanol subsidy outcome is monopoly power. Subsidies benefit an inconceivably unwise choice for energy – a low-energy, high-cost and highly ecologically destructive source – corn. These policies obstructed investment in truly green energy sources that are clearly scalable – solar, wind, waves, tides, geothermal and other renewable technologies. Had Congress put 20% of the ethanol subsidies into truly renewable energy sources, the US would be energy independent today. Instead, the current POTUS supports coal and ethanol.

Energy policy subsidies include extra payments consumers must make for **corn ethanol add $40 billion** a year to the true cost of fossil foods.

Hidden natural resource consumption

Deceit 6. Hidden natural resource consumption costs allow farmers to extract and over-consume fossil resources. Farmers should have a duty to preserve natural resources for future generations. Unfortunately, when future generations need vital resources such as fertile soil, fresh water and phosphorus, they will be extinct.

This situation, called the **tragedy of the commons,** occurs when a community relies on a shared-resource, e.g. fertile soil, clean air or clean water. The tragedy occurs when individual users act in their own self-interest and behave contrary to the common good. Individual users of the commons deplete or spoil the resource through their actions.

What do Fossil Foods really Cost?

A solution to the tragedy of the commons is to apply a tax on the common good. Oil companies pay an extraction fee, often 17% of the petroleum value. This reimburses the public for the exploitation of their natural resource. Farmers exploit and consume valuable fossil resources with no need for conservation because their extraction fees are zero.

IMA systemically degrades and destroys a resource more valuable than oil – soil. Farms in the US lose soil 20 times faster than the natural replenishment rate. Over the last 40 years over 30% of the world's cropland has become unproductive and abandoned due to erosion, salt invasion or nutrient exhaustion.[578] About 60% of the eroded soil flows into waterways, where turbidity fouls the water, sediment increases flooding risk and fertilizers and pesticides intensify water contamination.

Farmers know that row crops like corn and soybeans accelerate soil erosion from wind and water. They are aware that repeated planting of the same monocrop extracts nutrients and humus and accelerates soil exhaustion. They know too that one spring storm over cultivated land can carry off 40 tons of topsoil per acre. Incredibly, in spite of this knowledge, farmers plant more monocrop acreage, not less.

When repeated cropping wears out the soil or the topsoil has eroded, farmers must abandon their farm. Farmers may have lost their farm, but society has lost precious cropland, which is not economically repairable.

Water has become increasingly scarce for farms across the US. Hotter summer temperatures amplify water use due to increased evaporation and transpiration. Farmers often use irrigation water inefficiently. Open canals lose water from evaporation, theft and seepage. Farmers lose 50% of the remaining water to evaporation with flood or sprinkler irrigation, which are the most common application methods.

Many critical aquifers on which farmers, rural communities and cities depend will run dry in one or two generations. Farmers pay zero for constant over-draft that lowers the water table and causes:

What do Fossil Foods really Cost?

- Reservoirs, lakes and streams to drain.
- Land subsidence, which collapses buildings, roads, bridges and canals into sinkholes.
- Hydrology problems from deeper wells that increase pumping costs for all farmers and rural communities.
- Heavy metal invasion in deep well-water – lead, cadmium and arsenic – that poisons water for irrigation or household use.
- Accelerated aquifer crash.

Water for agriculture is not economically replaceable. Desal is far too expensive for fields. Expensive ocean water would have to be pumped up-hill, over the Coastal Mountain Range. When the aquifers go dry in California and the Mid-West, IMA fossil farming will go extinct.

Knowledge does not seem to change farmer behavior. The farmer mantra seems to be: *I will get my mine before someone else does*. This greed creates a run on water and the ruinous waste of our precious resource.

Natural resource consumption is a public subsidy to fossil food producers. A tax of only 10% of the fair cost for their consumption of public resources would probably double the cost of fossil food. Society would benefit from consumption taxes because farmers would use public resources more wisely.

Hidden natural resource consumption adds $50 billion a year to the true cost of fossil foods.

Deceit 7. Hidden environmental impact costs allow farmers to waste and pollute fossil resources without consequence. IMA farmers benefit significantly from these hidden unpaid costs, which are called externalities. Farmers get a free pass on externalities that take a terrible toll on people, animals, croplands, waterways and ecosystems.

Agriculture, including food processing and transportation, pays nothing for their 40% contribution to total GHG emissions that drive global climate chaos.

What do Fossil Foods really Cost?

What could be worse than allowing farmers to pay nothing for externalities caused by their eco-insensitive behavior? The only thing worse is creating a set of extremely irrational subsidies that give farmers incentives for farming practices that severely damage human and environmental health. Consider these rationalizations.

- I know that my crops carry hidden hunger that drives an epidemic of obesity, diabetes, CHD, cancers and other Western diseases. Why should I pay the extra costs to apply the micronutrients? They cannot blame me. I am only taking the incentives the USDA, Farm Bill, and Farm Bureau subsidies give me.

- I know monocultures destroy soil fertility, biodiversity and ecosystems. Monocultures put the entire food supply at risk from a single temperature spike or one pest vector. I will maximize wealth while these ludicrous subsidies allow me to profit.

- I know my eroded dust and overspill are poison carriers that pollute and degrade air, waterways and ecosystems. Generous subsidies force me to farm the duoculture of corn and soybeans.

- I know that GMO monocultures require more water, cultivation, herbicides, pesticides and fungicides than non-GMO crops.

- I know that my farm effluents, waste and pollution fouls air, waterways and ecosystems. I have no incentive to reduce emissions from animals, fertilizer, diesel equipment, pesticides or erosion. I will maximize my wealth while these senseless subsidies allow me to waste and pollute."

Clean-up of the Gulf of Mexico dead zone has been estimated to cost $2.7 billion a year. Cleaning thousands of miles of fouled US waterways would cost well over $200 billion a year. Cleaning air from agriculture GHG, nitrogen volatization and black soot particulates would cost over $200 billion a year. Restoring eroded and worn out cropland would cost trillions. Ag-degraded ecosystem restoration would cost more trillions, if restoration were possible.

Hidden **environmental impact costs add at least $100 billion** to the true cost of fossil foods.

What do Fossil Foods really Cost?

Full cost of fossil foods

Cost accounting for hidden fossil foods costs shows how truly expensive modern foods are to our economy and society. These hidden costs should be added to the prices of IMA commodity products.

Cost Accounting for Fossil Foods	Annual hidden costs
1. Hidden hunger inflicts medical costs	$1,100 billion
2. Hidden farm cash subsidies	$40 billion
3. Hidden executive orders	$3 billion
4. Hidden non-cash farm subsidies	$50 billion
5. Hidden energy policy subsidies	$40 billion
6. Hidden natural resource consumption	$100 billion
7. Hidden environmental impact costs	$100 billion
Total annual fossil foods hidden costs = $1,433 billion	

The hidden costs in freedom foods = zero. Freedom foods do not need subsidies and will reduce the health costs that fossil foods inflict.

The true cost of food should be transparent, not hidden. The only reason to deceive the public and hide the true cost of fossil foods appears to be that the USDA wants to continue its unprecedented wealth transfer to Big Ag wealthy white men without public scrutiny. If the USDA continues with the policy of subsidizing fossil foods that harm to their own children, the USDA should publish their rationale. Their reasoning will make unpleasant reading for their children.

USDA and prejudice

Why are 99% all Iowa farmers white men? The USDA's long history of racism is responsible. EEOC laws forbid discrimination and prejudice in American business and society, but not all abide.[579]

The USDA Commission on Small Farms admitted in Pigford v. Glickman, (1998) that "the history of discrimination at the USDA...is well-documented."[580] Farmers rely on loans and crop insurance, which

What do Fossil Foods really Cost?

the USDA is chartered to provide. It seems that pattern and practice favored only white men, which explains why 92% of US farmers are white men.[581] President Obama announced a $1.25 billion settlement, Pigford II in 2010 to fund some minority farmer claims.

1.   Current USDA actions appear to be prejudiced against other than senior white men, which is against American norms and laws.

If the USDA has a rationale and defense for their action, they have not published their defense. Every industry should follow American laws.

Subsidy payments of over $60 billion a year raises another question.

2.   If the USDA pays more than $60 billion a year in farm subsidies to nearly 100% wealthy white men, are those subsidies prejudiced and discriminatory behaviors?

Taxpayers are not aware of the hidden fossil food costs because they are purposely concealed. Are taxpayers willing to continue to fund farm subsidies when tens of billions of dollars go to white men who are already million and billionaires?

Fossil foods prejudice is the belief in the superiority of one food production system over another, which results in discrimination. This practice amplifies the existing racial prejudice in fossil food production.

3.   The USDA practices fossil foods prejudice. Fossil foods receive 100% of farm subsidies, the USDA-favored production method.

Zero USDA farm subsidy dollars go to alternative food production methods such as organic, resilient or abundance. Over 99% of USDA R&D dollars go to IMA, with <1% to organic methods.

The USDA intentionally obstructs innovation in food production with pattern and practice actions prejudiced for fossil foods. Obstructing benefits for human health appear to be a violation of public trust and certainly counter to the USDA Congressional mandates.

America and our world need innovation and diversity in our food supply chain, not deliberate obstruction that impede new methods.

What do Fossil Foods really Cost?

Summary

The cost of food should be transparent to consumers. The USDA and Big Ag deceits cause shocking harm to American consumers, farmers, natural resource stores and the health of our planet.

Those hidden direct and indirect subsidies of $1,433 billion a year provide incentives for farmers to consume, waste and pollute. These behaviors are exactly the opposite of the original farm subsidies' intent. Do subsidies use taxpayer dollars in the way they are intended?

If the subsidy boondoggle were put to the vote of the American public, less than 10% would support farm subsidies. Subsidies cost far too much and motivate the wrong behaviors. Farm subsidies are problematic. Subsidies promote only fossil foods, severely damage human health and purposely obstruct innovation and healthier food production methods.

The next US Secretary of Agriculture will receive awkward policy questions about the phenomenal cost of supporting IMA and fossil foods when the abysmal outcomes are well-known. Prior Ag Secretaries did not have to answer tough questions because fossil foods had no challenger. Those days are past. Aeroponics, resilience, organic, abundance and several other methods provide superior alternatives. The debate needs to begin now.

The desperately needed innovation for new food supplies are not likely to come from the USDA. The advance for healthier foods may come from the diffusion of abundance methods with the Emerald Bridge.

18. Diffusing Freedom Foods – The Emerald Bridge

The Emerald Bridge provides a move towards better health and food justice, where everyone has access to the substantial benefits of healthier freedom foods.

Key takeaway:

- How might we share the benefits of abundance methods?

The Emerald Bridge offers multiple paths to restore health for the over 280 million Americans suffering from micronutrient deficiencies caused by their diet of fossil foods. The Emerald Bridge proposes demonstration projects, closely linked to communities, that monitor key food consumption and health metrics.

Our path forward will use the R3D model: research, development, demonstration and diffusion. Biotechnologists and universities have designed productive biosystems and grower support technologies.

My prior books including *Abundance, Smartcultures, Freedom Foods* and *Peace Microfarms* have described the R&D and demonstration biosystems that use abundance methods to produce freedom foods.[582] Robert Henrikson's *Algae Microfarms* provides an excellent guide for microfarmers.[583] The focus here will be on pilot demonstration, diffusion and consumer adoption.

Diffusing Freedom Foods – The Emerald Bridge

The Emerald Bridge model below provides a means for diffusing the Emerald Renaissance with its substantial value proposition. Pilot projects will test the degree to which users benefit from freedom foods.

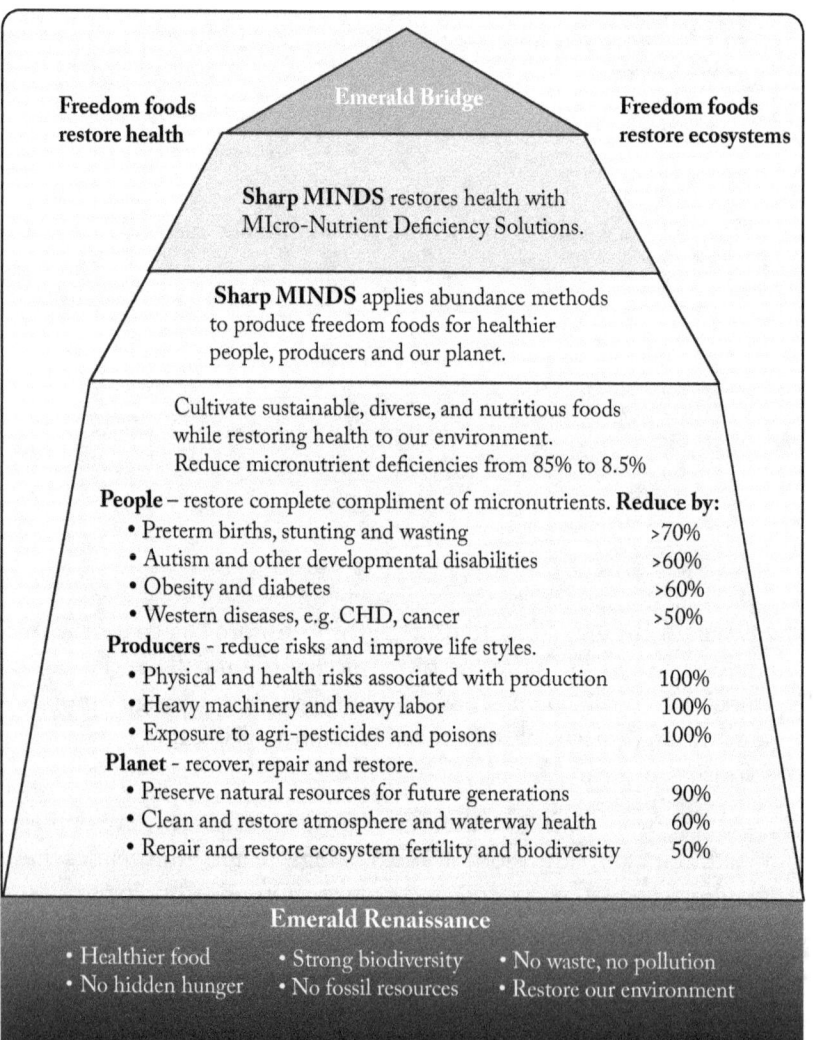

Freedom foods restore health

Emerald Bridge

Freedom foods restore ecosystems

Sharp MINDS restores health with MIcro-Nutrient Deficiency Solutions.

Sharp MINDS applies abundance methods to produce freedom foods for healthier people, producers and our planet.

Cultivate sustainable, diverse, and nutritious foods while restoring health to our environment.
Reduce micronutrient deficiencies from 85% to 8.5%

People – restore complete compliment of micronutrients. **Reduce by:**
- Preterm births, stunting and wasting — >70%
- Autism and other developmental disabilities — >60%
- Obesity and diabetes — >60%
- Western diseases, e.g. CHD, cancer — >50%

Producers - reduce risks and improve life styles.
- Physical and health risks associated with production — 100%
- Heavy machinery and heavy labor — 100%
- Exposure to agri-pesticides and poisons — 100%

Planet - recover, repair and restore.
- Preserve natural resources for future generations — 90%
- Clean and restore atmosphere and waterway health — 60%
- Repair and restore ecosystem fertility and biodiversity — 50%

Emerald Renaissance

- Healthier food
- No hidden hunger
- Strong biodiversity
- No fossil resources
- No waste, no pollution
- Restore our environment

Diffusing Freedom Foods – The Emerald Bridge

Emerald Bridge goals

The Emerald Bridge, a restorative health-bridge to communities, will deliver **Sharp MINDS,** MIcro-Nutrient Deficiency Solutions.

The Emerald Bridge drives three primary goals, **Sharp Minds:**

- Reduce the current 85% nutrient-deficient Americans to <8.5%.
- Cut pre-term births by 70%.
- Reduce developmental disabilities by half, to 7% from 15%.

Emerald Bridge goals are segmented in terms of improving health for people, producers and our planet.

Human health

Over 85% of Americans are nutrient deficient due to their fossil food diet. Nutrient deficiencies cause preterm births, birth defects, learning disabilities, mental retardation, reduced immunity, blindness, weak brains that result in poor learning, attention deficits and work capacity. Nutrient deficiencies also result in the litany of diseases associated with the Western Diet - CHD, cancers, neurological disorders, allergy and respiratory diseases and premature disability and death. [584]

Sharp Minds starts with the root-cause analysis for micronutrient deficiencies. Fortunately, the EAT-Lancet report and other sources[585] identify nutrient deficiency as the root-cause. The Western Diet of fossil foods that contain hidden hunger and empty calories.[586]

Smart Minds will provide freedom foods to people to restore their nutrient health. The therapeutic mechanism behind the ability of freedom foods to restore nutritional health is straight forward:

When nutritionally deficient consumers eat high-nutralence freedom foods, their nutrition status returns to normal.

Considerable research already exists that shows directly through microcrop supplement trials or indirectly from epidemiological studies.[587] Freedom foods such as spirulina, chlorella, sea vegetables or other microcrops, resolve micronutrient deficiencies.[588] These studies have not been done in the US.

A survey of 100+ scientists that work with microcrops will assess their confidence that microcrops will repair micronutrient deficiencies. While short of demonstrated change, a survey will validate the science behind the Smart Minds project.

Proof of Concept

Proof of concept will include several steps orchestrated by an Emerald Design Team of medical, science, nutrition and social scientists.

1. Identify several communities with known high micronutrient deficiency rates that benefits from the presence of a college or community college. The local trial project will be coordinated with the local academic and social community.

2. Survey communities to assess willingness to participate in a 12-week program. Participation will include only a sample of community members. The community agreement includes:

   a. Three blood tests, pre-, 6-weeks and 12-weeks, to assess micronutrient levels. Eventually, a non-invasive nutrient hand-scanner will serve the metrics tracking function.

   b. Willingness to eat 5-ounces of freedom food twice a day, that may be delivered in a breakfast bar, powdered drink or snacks.

3. Connect with community groups such as Feed America, Food Banks, churches and March of Dimes to plan food distribution.

4. All individual data will remain confidential. The Emerald Team will determine data analysis and public reports.

In the initial phase, freedom food will be sent to each community for distribution. Later projects will include microfarm construction to grow freedom foods for the community. Microfarmers will receive training on how to grow healthy microcrops. Microfarmer will commit to 3-hours of cultivation and harvesting per day. The project will pay for cultivation input costs.

Diffusing Freedom Foods – The Emerald Bridge

The Emerald Bridge Phase II and beyond will be planned by the Emerald Design Team and partners that emerge. The goal to reduce micronutrient deficiencies by a factor of 10 will require building a social and economic network. The network will help meet the goals of social, health and food justice, are where everyone has access to healthy food, and no one suffers from micronutrient deficiencies.

Additional Emerald Bridge projects may focus on Strong Babies, ending preterm births, and Strong Children, addressing obesity and diabetes. Microcrops deliver rich bioactive compounds not present in terrestrial food sources that address medical issues for nearly every human organ. Bioactive compounds offer Emerald Bridge projects emphasizing strong: brain, hearts, lungs, vision and digestion. Other possibilities are avoidance of CHD, cancer, auto-immune diseases and dementia.

Producer health

IMA farmers face many substantial risks – health, economics and social. Farm families, farm animals and rural communities must endure considerable health risks from IMA waste streams and pollution.

The best way to demonstrate avoidance of these risks is simple: survey producing microfarms and assess microfarmer risk. Microfarms are designed to cultivate superior food while minimizing producer risk by at least a factor of 10. Avoidance of big equipment, heavy labor and agri-chemicals and poisons reduce grower risk substantially.

Planet health

Our planet's health endures multiple insults from IMA farming practice. The Emerald Bridge project will refine the metrics for IMA methods for comparison with abundance methods used in operating microfarms.

A valuable opportunity for planet health assessment is the creation of a natural resource accounting system. Accounting for natural resources will attach a value on each resource, which makes quantifying resource preservation visual and valuable.

Natural resource accounting has not been applied to IMA. Placing a value on natural resources will spur the public to charge farmers for strip-mining cropland, allowing massive topsoil erosion, wasting fossil water and extracting agri-chemicals. Resource accounting will also clarify the damage IMA agriculture imposes with its massive waste and pollution.

Change metrics

Four Emerald Bridge actions to quantify IMA damages and serve as change agents that create a public demand for a transformation of our food supply system. Quantify IMA's 4P health insults:

1. **People** – linked diseases and medical conditions.
2. **Producers** – farmers and farm families deserve better lives.
3. **Preservation of natural resources** – extraction and consumption.
4. **Our planet** – the enormous and destructive waste and pollution.

IMA defenders may promise better behavior. The probability that IMA practice in any of the 4P areas can improve by even 5% approaches zero. The gap delta between IMA and abundance on the 4P health dimensions hover around 90% – at least two orders of magnitude.

Gate keepers

Four pivotal groups act as gate keepers. Each needs to be convinced that freedom foods can be produced efficiently, improve health and are attractive and affordable. Gate keepers demand to understand how freedom foods can prevent and reverse the terrible damage fossil foods have inflicted on society, which is why change metrics are critical.

1. **Consumers** will only make diet changes if they can see that one choice benefits them *on their terms* more than another. Consumer behavior is fickle and idiosyncratic. Some consumers are personal or family-health centric. Others may be persuaded by benefits to producers, preservation of resources or restoring our environment.

2. **Producers** will make production changes only if they know they have a market and are able to profit from production. A diverse cadre of Green Masterminds are eagerly waiting for good jobs growing freedom food in cities, towns and communities of all sizes.

3. **Food companies** will add to their marketing mix only if they are assured the market, (sufficient consumers) will make smart purchase decisions because the health benefits are compelling.

4. **Health professionals** will make dietary recommendations only if they have scientific proof that new foods provide protection against micronutrient deficiencies, Western diseases and therapeutic solutions for people with major medical conditions.[589]

Farm Bill – No!

Integrating the Emerald Bridge into the Farm Bill should be a logical track. Unfortunately, this strategy is a non-starter. The Farm Bill is renewed every five years, which is far too slow. Mega agri-businesses control Congress, agri-funding and especially the language used in the Farm Bill. Years of agri-policy legislation have failed to decrease fossil food 4P health insults. Instead agri-policy increases health abuses.

Big Ag and the USDA have shown little commitment to assure healthy consumers, farmers, ecosystems or sustainable food. The USDA subsidy behaviors show that their policy maneuvers benefit only Big Ag.

Consumer adoption

The health metrics goals in Emerald Bridge graphic are compelling. The goals offer substantial value in conveying freedom food health benefits. Metrics are critically important because they provide a visual measure of progress towards those goals. Goals such as reducing preterm births by 70% are doable. When mothers that have the full set of micronutrients they need, they are far less likely to have early deliveries.

Imagine a community that implements an Emerald Bridge Sharp Minds project and demonstrates that freedom foods eliminate micronutrient deficiencies and the terrible maladies that they impose. Each Emerald Bridge project will serve as a link to new communities and new adopters.

Every Emerald Bridge venture will add to the growing big data set. The Framingham health study set the standard for health data that has guided health professional decisions for decades.[590] The human health metrics will create a bandwagon effect, where everyone wants to get on board and not be left to the well-known ailments inflicted by fossil foods.

Plant-based diet

Nearly all the global human health and nutrition studies, including *EAT-Lancet*, recommend a change from animal meat to a plant-based diet. A meta-study on red and processed meat found that 50-grams of processed meat a day increases the risk of colorectal cancer by 18%.[591]

The American Institute for Cancer Research recommends to "limit" red meat but "avoid" all processed meat.[592] Colorectal cancer causes over a million new cases and 600,000 deaths yearly.

Diet change will improve human health and reduce the ecological damage. However, decades of consumer behavior research show that few eaters are willing give up their favorite meats easily.

The Emerald Bridge solution recognizes consumers have negative inertia to adopt something novel to them. We plan to design healthy microcrop-based meats and dairy products of every type, taste and texture that exceed meat eaters' expectations. Plant-based meats and dairy products must meet or exceed consumer pallet expectations before buyers will choose them.

Consumers are beginning to adopt plant-based foods as 2018 sales exceeded $5 billion.[593] Tyson will begin making *great tasting protein alternatives that are more accessible for everyone.* Investors and big food conglomerates are placing large bets on plant-based foods, especially meat. Cargill, Tyson Foods and General Mills have each made substantial investments in plant-based protein startups recently.

Acquisition targets include Puris Proteins, Memphis Meats, Beyond Meat and Super Meat. Beyond Meat plans an initial public offering in Q2 of 2019. Danone acquired WhiteWave Foods, which plans to triple its plant-based product sales, from $1.9 billion to $5.7 billion by 2025.

Impossible Foods began its incredible journey with Impossible plant-based burger sales through White Castle.

Impossible burgers have upscaled and are available now at many high-end restaurants and grocery retailers, including Whole Foods. Impossible Foods launched its burgers in Hong Kong in 2018 and plans to expand throughout Asia.

Producer adoption

Microfarmers will include many new growers, and others who currently support community gardens or farmers' markets. Microfarmers will be attracted to the healthy lifestyle and contribution to their community. They will be supported by a collegial network of **Green Masterminds** that share their passion for cultivating freedom foods. The Green Masterminds network will provide remote monitoring and a help-line to assure growers have the answers they need. The Green Masterminds network operates similarly to the Fédération des Spiruliniers de France.

Over 100 algapreneurs, including Laurent Lecesve, (left) at his Eco-Domaine farm in Normandy produce algae locally today and over 500 are expected within five years.

The algae microfarm movement that began in France has migrated to Spain. The CFPPA Center in Hyères trains growers.

Industrial farmers will likely begin with SmartCultures biosystems that serve multiple purposes. Farmers with overspill problems can place microfarms at the end of their fields where water tends to run-off. They can use the algae biomass on their fields to reduce their fertilizer cost and to improve crop nutralence, color, size and yield. Smart farmers will use Nrich processes to systemically improve soil fertility and structure.

Politicians in Florida, Michigan and other states with severe problems with fertilizer-driven eutrophication will pass laws to end IMA pollution. These new laws may mandate either the use of microcrop-based fertilizer that have high absorption rates or microfarms that catch wastewater. Both solutions will benefit farmers and citizens immensely. Currently, citizens must endure ugly, smelly and deadly eutrophication.

Food companies

Smart food companies already offer a wide array of sea vegetable snacks and functional foods. Especially popular are protein drinks and bars containing whole algae or algae-based nutrients. Whole Foods, Safeway, Kroger's and others are building new product categories for microcrop-based snacks, salads, noodles, drinks and meats. Freedom foods will become a product extension available in every aisle.

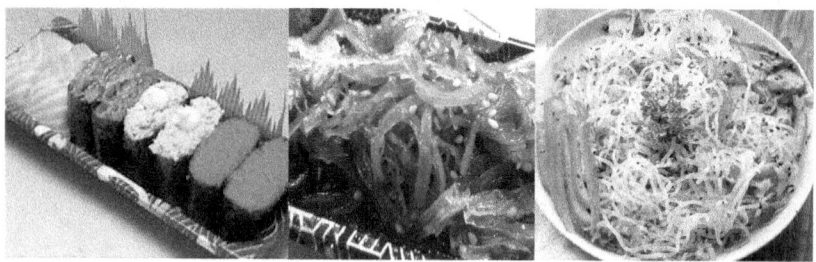

Sushi nori sea vegetable and kelp salads

Health professionals

Health metrics, empirical results from field studies, will be the key to persuade health professionals to recommend freedom foods.[594] Many doctors and nutritionists today recommend diets lower on the food chain. Algae resides at the bottom of the food chain, which creates many nutritional and health advantages, including especially high nutralence.

Health professionals value speed, the time between treatment and response. When people receive the full complement of micronutrients, health restoration for many maladies occurs quickly. When obese children eat foods that suppress the nosh response, those children quit over-eating. Positive health metrics will encourage early triers and early adopters to recommend freedom foods to their family and friends.

The medical literature aligns with the benefits provided today by many functional foods. Freedom foods will extend those benefits to regular foods such as microcrop-based bread, pasta, drinks and plant-meats.

Stop fossil foods?

No, farmers do not have to stop fossil food production. Farmers will continue using intensive mechanical methods to produce food – until the first of the 24 critical fossil resources becomes unaffordable in their geography. Expanding human populations will need every conceivable food source to supply sufficient food for the billions of hungry mouths expected to share the earth's limited resources by 2050.

Fossil foods will become legacy foods that remain available similar to gasoline and diesel cars. Some consumers may choose to pay a premium for the "good old taste of bad nutrition." Animal meat products may become the last of the legacy foods, although consumers will have to pay several times the cost of a microcrop burger to damage their health.

By 2025, some smart cities or states will impose eco-taxes on animal products based on the harm production does to the environment. Farmers will face huge cost increases as eco-taxes are imposed on extraction and consumption of public non-renewable resources. Fossil farmers will transform to freedom food production quickly when:

> *Public policy regulations begin fair taxes on farm extraction, consumption, waste and pollution.*

Abundance methods can grow animals for meat and dairy while using minimal fossil resources and creating zero net-pollution. However, abundance methods will focus on plant-based meats because the food conversion ratio, 1:1, is far superior for each of the 4P metrics.

Plant-based meats

The transition to plant-based meats will accelerate from drivers in each of the 4P health categories.

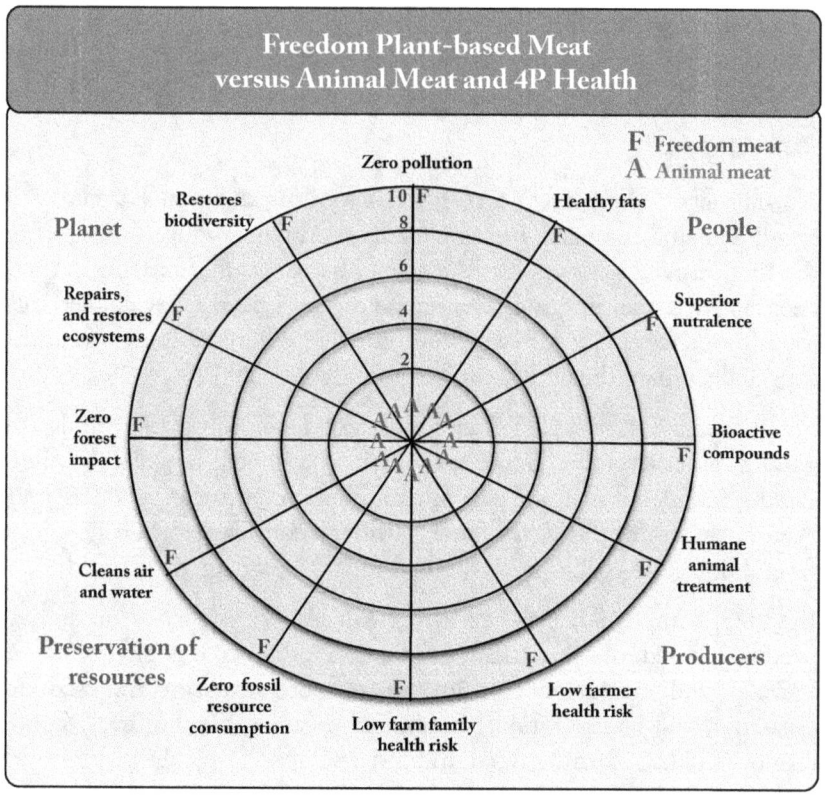

Freedom Plant-based Meat versus Animal Meat and 4P Health

Several target markets will be early plant-based meat adopters.

- *New York Times* food journalists for healthier nutrition.
- Sierra Club and Green Peace for saving our planet.
- Audubon Society for saving our birds, bats, butterflies and bees.
- 4H and Future Farmers of America for sustainable farm practice.

Legacy farmers

Farmers have been following USDA subsidy incentives for decades. That will change immediately when subsidies stop. Many farmers will rejoice in their new freedom and plant diverse crops that benefit instead of degrading consumer health. Legacy field crop farmers will find lucrative niche markets and others will add microcrops to their production mix to create multiproduct opportunities. However, legacy field crop farmers will not be able to compete with microcrops without roots. Microcrops can produce superior protein and other nutrients 50 times faster. IMA farmers cannot touch the 10x plus microcrop nutralence advantage, nor any of the 10 freedom food advantages.

Allowing free markets to sort out differences in production methods would impose an unhealthy red-ink balance-sheet revolution for millions of hard-working farmers, forcing bankruptcy. The USDA can help.

1. The USDA might shift current farm subsidies to farm buyouts. A buyout for the average 58+-year-old farmer might be the average net income from the prior four years, prorated for each year to age 62, when the farmer reaches Social Security eligibility. Buyouts might parallel the practice of buying water rights, which today takes many farmers out of fossil agriculture.

2. All farmers, whether they choose a buy-out or not, would have access to multi-modal Emerald Renaissance training in abundance methods and the production of freedom foods.

Trainings will cover how to repair the farmer's soil and ecosystems with abundance methods. If the farmer chooses to grow field crops, the trainings would show how to use microfarms to biocycle nutrients, prevent overspill, minimize erosion and a host of actions that reduce farming costs and health risk and improve sustainable food production. These training will integrate organic and resilient methods.

The Farm Bureau, FFA, 4H, Iowa State University, Cornell, UC Davis and other competent agribusiness sources will come forward with solutions to help IMA farmers transition to healthier food production methods. The most important assistance to farm families may be adopting new food production methods that reduce risk to farmers, their families and their communities.

Diffusing Freedom Foods – The Emerald Bridge

Emerald Bridge task

The Emerald Bridge primary goal, to provide every American with sufficient micronutrients, does not require changing the entire food supply chain. Resolving micronutrient deficiencies requires only about 5 grams of a high-nutralence microcrop like Spirulina daily. The packet of snack flakes (right) shows this is only about a tablespoon.

High-nutralence freedom food supplements can be made into any form: shakes, flakes, chips, cookies, bars, nuts or snacks. Freedom foods can be made into nearly any form that consumers are likely to prefer for breakfast, lunch or dinner. Forms include microcrop-based flour, oil, plant sugar and plant-based meats. Home cooks, chefs and food formulators can make just about any food consumers prefer today, free from the health anxiety associated with fossil foods.

Nutritionists will not have to throw away the food pyramid because most choices can stay in place. Meat and dairy may add an important moniker – plant based. Nutralence will allow opportunity for substantial additional nutritional information when measures for nutrient quality, density, diversity, bioavailability and bioactive compounds are available.

Funding

Funding should reflect costs, and fossil foods costs are shocking. IMA production, processing and transportation create substantially more cost than any other sector of American industry. Those costs include human, animal and environmental health, loss of natural resources and pollution. The petroleum industry is a far distant second on imposing social costs.

If the CDC or other agency added all the health costs fossil foods impose on society, only 1% of those costs would fund a national Emerald Bridge project to deliver freedom foods to every American within a decade.

If the USDA calculated all the costs fossil foods exact on farmers, farm families and rural communities, only 1% of those costs would fund a national project to deliver freedom foods to every American in 5 years.If the EPA created a summation of all the externalities fossil foods inflict on the environment, only 1% of those costs would fund a national project to deliver freedom foods to every American by 2030.

Targeted initiative

An alternative targeted funding policy might move forward more quickly. All societies are especially fond of their children. The plight of unborn children in America is ominous, similar to a 3rd world country.

> Preterm birth is the most frequent cause of infant death in the US, and costs the healthcare system $26 billion a year.[595]

Preterm deliveries cause underweight births, wasting and stunting and developmental disorders such as Autism. Only 10% of those costs would fund the Emerald Life initiative and save over 10,000 families a year from the perils of preterm birth. This funding would lead to the transformation to a healthy food system in 10 years. Only 1% of present costs could relieve an entire state from the terrible burden of preterm births and all the medical and family costs in 5 years.

Emerald Bridge diffusion

Strategy: Apply 10% of the annual cost of preterm infants, $2.6 billion, to produce peace microfarms for placement across the country.

1. Allot $2 B to pay for 20,000 microfarms, (at $100,000 each)
2. And $0.6 B for training growers and creating a smart network for microfarmers.

Microfarm distribution would be based on a qualified lottery of willing people in areas with high concentrations of micronutrient deficiencies and preterm births. Diverse growers would receive training through the AlgaeFoundation.org, community colleges, YouTube and other social media. Santa Fe Community College and collaborators ASU and UT Austin have developed an excellent short course for microfarmers. Each microfarmer would commit to giving half the freedom food to project participants. The March of Dimes, church networks and Feed America may assist with food distribution. The second half of each microfarmers' food could be sold locally.

Robert Henrikson's metrics from several years of Smart Microfarm production experience demonstrates that sales of the second half of the biomass would yield $15,000 to $25,000 a year for each microfarmer.

Vertical farm   *Rooftop microfarm*

The Emerald Bridge project would provide high social and medical value with a payback in under five years. An Emerald Bridge extension could reduce the stark obesity and diabetes problem by 50%. Emerald Bridge extensions could substantially reduce the occurrence of autism, ADHD, cleft pallet and a broad spectrum of maladies that occur from micronutrient deficits.

An Emerald Bridge project might place peace microfarms in prisons. Detainees could cultivate healthful freedom foods for their mates and their community. Peace microfarms would provide an excellent opportunity for training in sustainable freedom food production and the culinary arts. Prison-based microfarms offer a 4-year payback based only on internal food cultivation and waste management.

Peace microfarms can be designed for handicapped growers. This model might be staffed by disabled veterans. Since 1 in 6 US children are now born with developmental disabilities, microfarms might be staffed with people with disabilities. Many of these disabled children suffer their disabilities because their mother did not have access to sufficient micronutrients.

19. Can the Emerald Renaissance Transform our World?

The Emerald Renaissance proposes that abundance, a biologically-based novel food production system, promises to:

- Improve life quality and health for billions globally.
- Save >10 million lives a year in the US annually.
- Save >500 million lives a year globally.
- End or moderate the epidemics of obesity and diabetes.
- Substantially reduce early disability and death from major organ diseases; heart, brain, lungs, nerves, endocrine and skin.
- Prevent millions of premature disabilities and death, saving over 100 million premature life-years lost annually.
- Restore our diminished and degraded environment.

The Emerald Renaissance applies biological solutions and abundance methods to cultivate sustainable freedom foods that are healthier for people, producers, preservation of resources and our planet.

How much is the Emerald Renaissance transformation worth? The answer to this macro question requires a breakdown of each social, economic and environmental area. The sustainability Triple Bottom Line graphic on the next page provides the categories for value-added and savings. This model provides a roadmap for a fascinating doctoral dissertation, but it is too granular for adequate analysis here.

The Triple Bottom Line illustrates the direct value for a new form of food production in each area. Categories do not reflect the substantial savings from the avoidance of fossil resources, waste or pollution.

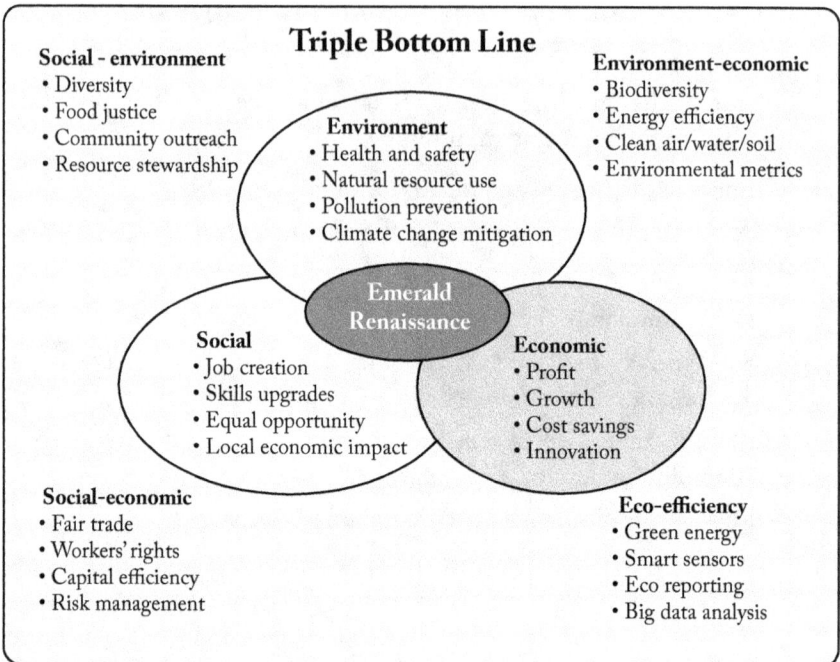

Triple Bottom Line

Social - environment
- Diversity
- Food justice
- Community outreach
- Resource stewardship

Environment
- Health and safety
- Natural resource use
- Pollution prevention
- Climate change mitigation

Environment-economic
- Biodiversity
- Energy efficiency
- Clean air/water/soil
- Environmental metrics

Emerald Renaissance

Social
- Job creation
- Skills upgrades
- Equal opportunity
- Local economic impact

Economic
- Profit
- Growth
- Cost savings
- Innovation

Social-economic
- Fair trade
- Workers' rights
- Capital efficiency
- Risk management

Eco-efficiency
- Green energy
- Smart sensors
- Eco reporting
- Big data analysis

Valuing the Emerald Renaissance

An alternative valuation approach examines the areas covered in prior chapters and sum the topline items. If you disagree with the valuation, please take out a pencil and make your corrections. The purpose here is not precision, but broad constructs for others to reflect and refine.

Heathier nutrition is worth more than all other Emerald Renaissance benefits combined. Sustainable food only has value if our grandparents, parents and children do not have their lives prematurely diminished, disabled or terminated by their fossil foods diet. All the ecological benefits of saving natural resources, avoiding pollution and repairing and restoring our environment become moot. People suffering the horrible pain of food-related health conditions completely miss the benefits.

Can the Emerald Renaissance Transform our World?

Food policy makers have an important choice – support foods that harm or foods that help and heal. Unbelievably, current US food policy supports foods that are well-known to cause catastrophic harm.

Fossil foods are directly implicated in nine of the 10 leading causes of death in the US.[596] Available metrics indicate the premature disability and deaths that occur annually from maladies imposed by fossil foods exceeds all deaths from drug and alcohol abuse, HIV/AIDS, traffic collisions, guns, mass shootings, terrorist attacks and war.

Ending premature disabilities and death from our malfunctioning food supply should be our top national priority. Improving human health while cleaning ecosystems and mitigating climate chaos adds to the strategic benefits.

The annual value for healthier people consuming freedom foods assumes that one-third of current US annual medical costs are food related, $1.1 trillion. Medical scientists may refine this value, probably upwards. If food-related medical costs are not mitigated quickly, these costs will not only destroy families and communities, they will destroy our society.

People - healthier foods that avoid micronutrient deficiencies, hidden hunger and Western diseases.	**Annual Value** $1.1 trillion

- Restore 2 million IQ points and avoid 7,500 intellectual disability cases that occur from the use of pesticides annually.597
- Cultivate foods that are cleaner, more nutritious and tastier, yet prevent rather than cause disease.
- Address social justice; where all people have access to good, affordable food.
- Prevent millions of preterm deliveries – low birthweight babies, stunting, wasting, autism and other developmental disabilities.
- Prevent and treat obesity and diabetes and many Western diet diseases, including heart, lung, brain, nerves and cancers.

Can the Emerald Renaissance Transform our World?

- Develop a library and monitoring system for nutralence – nutrient quality, density, diversity, bioavailability and bioactive compounds.
- Publish more research on microcrop disease avoidance, malady treatment and the value of bioactive compounds.
- End pesticide and agri-poison residuals on and in food.
- Stop food allergies and treat fossil food allergies.

Growers deserve lower risk, improved health and better lifestyles.

Producers – reduce economic and health risks for farmers, farm families and rural communities.	**Annual Value** $100 billion

- End the use once and done linear agriculture model.
- Reduce producer risk from crop failure.
- Eliminate use of monocultures, fossil fuels and inorganic fertilizers.
- Stop the severe health risks that accrue with agri-poisons.
- Enable growers to produce higher value multiproduct crops year-round independent of weather and climate chaos.
- Stop the systemic loss of topsoil from erosion and exhaustion.
- Eliminate accident and health risks from heavy machinery, fatigue, dust, sun exposure and black soot particulates.
- Abolish systemic air, water and ecosystem pollution that create extreme health risk for farmers, farm families and rural communities.

Current rates of natural resource extraction and consumption will crash stores in one or two generations. Fossil resources such as fertile soil, fresh water and phosphorus are not economically replaceable.

Preserve natural resources – save natural resources for our children and their children.	**Annual Value** $150 billion

- Embrace 10-zero foods.
- Move towards 4-zero production, manufacturing and distribution.

Can the Emerald Renaissance Transform our World?

- End the extraction and consumption of fossil resources for the entire food supply chain.
- Use renewable energy and an electric grid with sustainable storage.
- Impose extraction and consumption taxes on high-consumers of critical resources.
- Disclose resource consumption on food labels or websites.
- Publish names, individuals or companies, of the top 10% consumers of each vital natural resource.

Nature has been producing healthy foods successfully for eons without destroying herself. Natural ecosystems display scenic proof.

The Emerald Renaissance proposes that we learn from nature and repair and restore natural systems rather than raid and ruin our precious flora, fauna and natural landscapes.

Our Planet deserves healthier food production with less wear and tear and more tender loving care.	Value $200 billion

- End strip mining cropland and mend IMA scars.
- Terminate systemic air, water and ecological pollution and tax polluters. (Yes, some companies may have to pay both an extraction and a pollution tax, which is socially equitable.)
- Eliminate fertilizer and pesticide erosion and toxin migration.
- Abolish massive animal gas emissions and manure effluence.
- Bring exhausted and dead soil back to life.
- Restore biodiversity in flora and fauna, especially insects.
- Regenerate natural habitats and ecosystems.
- Capture and reuse carbon and other nutrients.
- Connect biosystems to pollution sources to biocycle nutrients.

Abolishing farm subsidies will save over $100 billion a year. Several annual hidden costs in the cost accounting for fossil foods are not included in this total since they factor into savings in other categories.

End farm subsidies - hidden farm cash subsidies, executive orders, non-cash farm subsidies and energy policy subsidies.	Annual Value $100 billion

Corn ethanol production greatly increases combustion emissions, which account for over 200,000 premature deaths per year and over 2 million life-years lost. Combustion emission deaths typically occur 10 years prematurely. Agricultural emissions are the leading cause of premature air pollution deaths in the US.[598] Death from air-pathway failure is gruesome. People fight perilously for every breath and then again for their last.

What has Big Ag, or the USDA done to lower ms? Nothing. Instead, the USDA subsidizes farmers to increase emissions.

Summary

The Emerald Renaissance proposes the adoption of a novel food production system that delivers healthier foods. New foods should be sustainable, use minimal or no fossil resources and put producers and farm family at no health or physical risk. These methods should produce foods with minimal or no waste or pollution. Several new methods show potential including hydroponics, aquaponics, aeroponics and abundance.

Abundance methods cultivate healthier foods and also repair and restore natural systems that have been degraded or destroyed by IMA. The annual saving from adopting abundance and other sustainable food production methods in America approaches $1,650 billion a year.

The absence of pesticides alone can restore 2 million lost IQ points and avoid 7,500 intellectual disability cases in newborns annually.

What other investment offers a similar return? Please lend your voice to support the Emerald Renaissance.

Acknowledgements

Special thanks to:

- John O'Hare, SCAD, Southern Cross Agricultural Development
- David and Marcia Pimentel, Profs. Emeritus, Cornell University
- Fred Krupp, President, the Environmental Defense Fund
- Ken Cook, President of the Environmental Working Group

Professors Qiang Hu, Chinese Academy of Science, Bruce Rittman and Klaus Lackner from Arizona State University.

Science	Business – Econ.	Agribusiness
Xuemei Bai	Mark Allen	Jon Ewen
Anastasia O'Rourke	Jim Niemann	Charles Green
Xun Wang	Maura Flight	Ben Cloud
David Punchard	Mark Ewen	David Kramer
Robert Henrikson	Mimi Hall	Doug Young
Stephen Mayfield	Rick Bellingham	Elham Fini
Amha Belay	Lieve Laurens	Rafael Quezada
Bobban Subhadra	Jim Hershauer	Barry Spiker

Also helpful were the published works of Paul Ehrlich, Sandra Postel, Nobel Laureate Al Gore, Harvey Blatt, Fred Pearce, Michael Pollen, Brian Halweil, Clay Jason and Linda Graham. High-content websites were a great support such as Algaebase, U.N., W.H.O., the National Resources Defense Council, Sierra Club, Green Peace, Audubon Society, Union of Concerned Scientists, Center for Energy and Climate Solutions, Drawdown, Clean Water Network and Public Citizen. Also useful were government sources DOE, EPA, NOAA and NREL.

Mark R. Edwards, Dr. Metrics

Mark passionately pursues one goal – resolving world hunger and health restoration for people, producers and our environment. Freedom foods provide healthier, sustainable food while simultaneously restoring clean air, water and ecosystems. He has written 20 books in the *Green Algae Strategy Series* that focus on sustainable food and energy.

Mark graduated from the U.S. Naval Academy, where he earned degrees in engineering, oceanography and meteorology. Jacques Cousteau motivated and mentored his interest in the oceans and global stewardship. He holds an MBA and PhD in strategic marketing and consumer behavior and taught agribusiness, innovation, sustainability, engineering, food marketing, leadership and entrepreneurship at ASU for 39 years. He served as marketing director for the Pritikin Longevity Center, where he helped design healthy foods and lifestyles. He served as director for a *Fortune 50* food company where he developed novel and healthier foods.

Mark founded and served as CEO of TEAMS Intl. for 22 years. He invented dozens of advanced metrics used by firms globally, including 360° feedback. He served as lead consultant for Disney, 3M, Monsanto, DuPont, Nabisco, Quaker Oats, General Mills, Borden, Coca-Cola, Frito-Lay, GE, Intel, J&J, HP, IBM, Merck, GM, BP, Glaxo and many other agribusiness, financial, health, transportation and technology companies. He has worked with leaders in over 12 countries.

Mark has published over 140 articles and 31 books that span business and science disciplines. His *360° Feedback,* with partner Ann Ewen, became a business best seller. Several science books won international best science and environment awards. Universities in over 30 countries use *Green Algae Strategy* series books in food, energy, economics and sustainability courses. He writes the popular *Algae 101* and *Algae Secrets* blog *for Algae Industry Magazine.*

Appendix I. The Green Algae Strategy Series

Mark R. Edwards

The Green Algae Strategy Series focuses on creating Sustainable and Affordable Food and Energy – "SAFE" production. **The Green Algae Strategy Series** are available for free downloading in color PDF at EmeraldRenaissance.com for students, teachers and food and energy policy leaders. They are also available on Amazon.com and other retailers. Teachers, professors and policy leaders use these books in schools and colleges globally for courses in sustainability, engineering, business, politics, social entrepreneurship, food, water, energy, ecology, environment and world future.

BioWar I: Battles over Food and Fuel Lead to World Hunger, 2008.

Green Algae Strategy: End Oil Imports and Engineer Sustainable Food and Fuel, 2008, Gold Medal, IPPY Best Science Book.

Green Solar Gardens: Algae's Promise to end Hunger, 2009.

Crash! The Demise of fossil Foods and the Rise of Abundance, 2009.

Smartcultures: Sustainable Food despite Climate Change, 2010.

The Tiny Plant the Saved our Planet, 2010. Nautilus Silver Medal, Best Children's Book.

Abundance: Sustainable Fossil-free Foods with superior Nutrition and Taste, 2010, Gold Medal, Best Environmental Book.

Abundant Agriculture: Redesigning a healthier food supply, 2011.

Freedom Foods: Superior Nutrition and Taste from low on the Food Chain, 2011, Gold Medal, Best Science Book.

Imagine our Algae Future: Results of the International Algae Competition, 2012. AlgaeCompetition.com.

Climate Independent Foods: Survive and Thrive Anywhere and in any Weather, 2012.

Peace Microfarms: A Green Algae Strategy to prevent War, 2012. Gold Medal, Best Science Book.

Green Friendship Bridge: Distributed Microfarms Slow Immigration and Create Social Justice, 2018.

Ana Feeds Our World by 2040: Miracles with Nature's Nano-cell Biofactory, 2018. Gold Medal, Best Science Book.

Emerald Renaissance: World Hunger Solutions Healthier for People, Producers and our Planet, 2019.

Fallout! The Dirty Bomb that Poisons our Food Supply, 2019.

Fieros: Taming the Fiercest Stealth Predator in History, 2019.

The Deadliest Decision in American History, 2019.

Appendix II. Algae's Amazing Miracles

Algae's actions over the past 3.5 billion years defies the imagination. She has created more than 20 times more miracles than any other organism on earth. Algae's family includes both single and multicellular organisms, which are classified as plants, bacteria, and by some scientists, as tiny animals with intelligence. Algae's extraordinary actions build a shared understanding for the miracles she will produce next.

Algae will improve our food, health, and medicines, while repairing many of the most degraded and polluted ecosystems on our planet.

Algae:

1. Became the first plant on the planet, around 3.5 billion years ago.
2. Survived and adapted to the brutal conditions of early earth.
3. Evolved photosynthetic ability to create biomass from sunlight.
4. Use photosynthesis to capture CO_2 and release O_2 to support life.
5. Became the foundation of the food chain for plants, then animals.
6. Became the mother of all land plants 500 million years ago.
7. Provides nutrients for rootless plants; moss, lichens and corals.
8. Learned to grow 30 – 100 times faster than any other organism.
9. Does not waste energy on superfluous roots, stems or leaves.
10. Learned to survive through multi-season dormancy.
11. Act altruistically; purposely dies so progeny have sufficient food.
12. Developed the highest nutralence of any organism.
13. Recovers carbon and other nutrients from gas and wastewater.
14. Recovers heavy metals from gas and wastewater.
15. Eliminates heavy metal poisons from human body tissues.

16. Improves stress tolerance and survivability for plants.
17. Activates and supports plants' natural pest defense systems.
18. Restores life to dead soil from nutrient extraction.
19. Restores soil dead from salt invasion, back to life and fertility.
20. Restores life, germination, to many dead or non-viable seeds.
21. Improves sperm count and motility for animals and humans.
22. Saves many animals from death with higher survivability rates.
23. Activates animals and human natural immunity defense systems.
24. Restores sight to blind mice and blind humans.
25. Protects cells from severe damage from free radical scavengers.

Future algae miracles

Algae are not done yet providing miracles. Algae promise to improve human societies and the earth with still more miracles. Algae will:

26. Provide sufficient food to ensure social justice, where everyone has access to affordable and healthy food.
27. Provide food justice, where everyone has access food production.
28. End food deserts with access to local healthy food.
29. Allow climate independent food production, anywhere on earth.
30. Enable freedom foods, grown with minimal fossil resources.
31. Provide nutrients to end malnutrition and nutrient deficiencies.
32. Restore 100 million IQ points for children with brain disorders.
33. Restore good health from heavy metals poisoning in children.
34. Provide food and life-support systems for deep space exploration.
35. Provide energy and medicines for space exploration.
36. Algae bioactive compounds protect from and provide treatments for autism, ADHD and other brain dysfunctions.
37. Health restoration from heavy metals and pesticide poisoning.
38. Obesity and diabetes with fibers and antidiabetic biocompounds.
39. Kill cancer cells with anticancer compounds and cancer toxins.

40. Heart failure and strokes with anticoagulants.
41. CHD with antihypertensive and antihyperlipidemic compounds.
42. Arthritis and hepatitis with anti-inflammatory compounds.
43. ALS, Parkinson's and other neurodegenerative diseases.
44. Asthma and allergies with immunomodulatory compounds.
45. Yeast and fungal infections, e.g. ringworm, with antifungals.
46. Asian flu, viral pneumonia and HIV/AIDS with antivirals.
47. Degenerative metabolic disorders.
48. Brain dysfunction including dementia and Alzheimer's disease.
49. Reduction in bipolar disorders and mental illnesses.
50. Substantial reduction in depression, PTSD, and suicide.

What an amazing tiny plant! She has not finished creating miracles. Please send your ideas, observations or discoveries of new miracles to EmeraldRenaissance.com.

References

1 https://www.sciencedaily.com/releases/2018/06/180603193616.htm
2 https://www.sciencedirect.com/science/article/abs/pii/S0006320718313636
3 https://eatforum.org/content/uploads/2019/01/EAT-Lancet_Commission_Summary_Report.pdf
4 https://espresso.economist.com/e40d53ca19cd28f7dae77368fab8df4d
5 https://www.iatp.org/documents/livestocks-contribution-15c-pathway-0
6 https://www.ncbi.nlm.nih.gov/pmc/articles/PMC3840875/
7 https://eatforum.org/eat-lancet-commission/
8 https://www.who.int/news-room/detail/11-10-2017-tenfold-increase-in-childhood-and-adolescent-obesity-in-four-decades-new-study-by-imperial-college-london-and-who
9 https://www.cdc.gov/nchs/fastats/obesity-overweight.htm
10 http://www.healthdata.org/news-release/vast-majority-american-adults-are-overweight-or-obese-and-weight-growing-problem-among
11 http://uconnruddcenter.org/files/Pdfs/TargetedMarketingReport2019.pdf
12 https://www.cdc.gov/chronicdisease/overview/index.htm
13 https://www.ncbi.nlm.nih.gov/pmc/articles/PMC4034518/
14 https://portal.nifa.usda.gov/web/crisprojectpages/0433776-food-markets--food-expenditures-and-marketing-costs.html
15 https://www.thelancet.com/gbd
16 http://www.fao.org/docrep/u8480e/U8480E07.htm
17 http://articles.latimes.com/2003/jun/15/nation/na-diabetes15
18 Seneff, Stephanie. Is Glyphosate the 'Safe' Chemical that's making us all Sick? Presentation on November 30, 2018 in Orlando, Florida, at the SOHO convention. *(Powerpoint)_(PDF)*
19 https://www.ers.usda.gov/topics/food-nutrition-assistance/food-security-in-the-us/key-statistics-graphics.aspx
20 Ibid.
21 http://www.cdc.gov/reproductivehealth/maternalinfanthealth/PretermBirth
22 Boyle CA, et al. Trends in the Prevalence of Developmental Disabilities in US Children, 1997–2008. Pediatrics. 2011; 27: 1034-1042.
23 http://connects.catalyst.harvard.edu/Profiles/display/Person/73533
24 Pimentel, D.; Satkiewicz, P. Malnutrition. In Encyclopedia of Sustainability, Volume Natural Resources and Sustainability; Berkshire Publishing Group: Great Barrington, MA, USA, 2013

[25] Qian Di, M.S. et al. Air Pollution and Mortality in the Medicare Population, New England Journal of Medicine, 2017; 376:2513-2522.

[26] http://www.ehsciences.org/

[27] Ibid

[28] https://www.ucsusa.org/food_and_agriculture/our-failing-food-system/industrial-agriculture/prescription-for-trouble.

[29] https://www.ewg.org/foodnews/dirty-dozen.php

[30] https://uspirg.org/feature/usp/glyphosate-pesticide-beer-and-wine

[31] http://www.panna.org/issues/publication/chemical-tresspass-english

[32] https://www.statista.com/statistics/306105/us-acreage-of-genetically-modified-soybeans/

[33] Charles M Benbrook. Impacts of genetically engineered crops on pesticide use in the U.S. -- the first sixteen years. *Environmental Sciences Europe*, 2012; 24 (1): 24

[34] Attina, Teresa M., et al. Exposure to endocrine-disrupting chemicals in the USA: a population-based disease burden and cost analysis, *The Lancet: Diabetes and Endocrinology*, Dec 01, 2016, 4:12, 996-1003.

[35] https://www.cdc.gov/nchs/products/databriefs/db328.htm

[36] https://news.nationalgeographic.com/news/2005/12/agriculture-food-crops-land/

[37] http://fortune.com/2016/10/27/5-industries-hurt-job/

[38] www.naturalnews.com/046193_GMO_crop_failures_farmer_suicides_India

[39] https://news.nationalgeographic.com/news/2012/05/120531-groundwater-depletion-may-accelerate-sea-level-rise/

[40] https://www.scientificamerican.com/article/time-to-rethink-corn/

[41] Lal, R. Soil Erosion and Land Degradation: The Global Risks. In Soil degradation; Lal, R., Stewart, B.A., Eds.; Springer-Verlag: New York, NY, USA, 1990; pp. 129–172.

[42] https://news.nationalgeographic.com/news/2014/05/140529-conservation-science-animals-species-endangered-extinction/

[43] https://www.sciencedirect.com/science/article/abs/pii/S0006320718313636

[44] Extinction risk from climate change, CD Thomas, et al., Nature 427 (6970), 145.

[45] https://www.epa.gov/ghgemissions/global-greenhouse-gas-emissions-data

[46] https://www.iatp.org/documents/livestocks-contribution-15c-pathway-0

[47] https://link.springer.com/article/10.1007/s00376-019-8276-x

[48] https://news.agu.org/press-release/groundwater-pumping-leads-to-sea-level-rise-cancels-out-effect-of-dams/

[49] https://www.pnas.org/content/early/2019/01/08/1812883116

[50] Diaz, R. J., & Rosenberg, R. (2008). Spreading Dead Zones and Consequences for Marine Ecosystems. Science, 321(5891), 926-929.

[51] http://www.onegreenplanet.org/animalsandnature/facts-on-animal-farming-and-the-environment/

[52] http://edis.ifas.ufl.edu/pi180

[53] https://www.amazon.com/Oceana-Endangered-Oceans-What-Save/dp/B005Q5OYHY https://www.amazon.com/Oceana-Endangered-Oceans-What-Save/dp/B005Q5OYHY

[54] http://www.fao.org/docrep/016/i3010e/i3010e01.pdf

[55] http://www.fao.org/docrep/007/y5609e/y5609e02.htm

[56] https://www.ipcc.ch/sr15/

[57] https://www.ipcc.ch/sr15/

[58] https://www.thelancet.com/journals/lancet/article/PIIS0140-6736(18)31788-4/fulltext?utm_campaign=tleat19&utm_source=hub_page

[59] Lal, R. Soil Erosion and Land Degradation: The Global Risks. In Soil degradation; Lal, R., Stewart, B.A., Eds.; Springer-Verlag: New York, NY, USA, 1990; pp. 129–172.

[60] Pimentel, D.; Pimentel, M. Food, Energy, and Society; CRC Press: Boca Raton, FL, USA, 2008.

[61] http://thescienceexplorer.com/nature/why-we-need-biodiversity-our-crops

[62] http://www.fao.org/docrep/u8480e/U8480E07.htm

[63] Patricio Grassini, Kent M. Eskridge, Distinguishing between yield advances and yield plateaus in historical crop production trends, *Nature Communications*, 4, Article number: 2918 (2013).

[64] Ibid.

[65] https://www.treehugger.com/corporate-responsibility/shifting-the-agricultural-research-paradigm-green-revolution-ii.html

[66] https://www.worldhunger.org/world-hunger-and-poverty-facts-and-statistics/

[67] E. D. Perry et al. Genetically engineered crops and pesticide use in U.S. maize and soybeans, *Science Advances* (2016). **DOI:1126/sciadv.1600850**

[68] https://phys.org/news/2012-10-superweeds-linked-herbicide-gm-crops.html#jCp

[69] https://bmjopen.bmj.com/content/6/3/e009892.full

[70] Loopstra, R, et al. Food insecurity and social protection in Europe: quasi-natural experiment of Europe's great recessions 2004–2012, *Preventive Medicine*, 2016, 89:44, 50.

[71] WHO. Nutrition for Health and Development: A Global Agenda for Combating Malnutrition; Progress Report Rome, Italy, 2000.

[72] FAO. FAO Food Balance Sheets. FAOSTAT, Food and Agriculture Organization of the United Nations, 2004.

[73] https://globalnutritionreport.org/reports/global-nutrition-report-2018/executive-summary/

[74] https://glopan.org/sites/default/files/pictures/CostOfMalnutrition.pdf

[75] https://www.who.int/nutrition/publications/foodsecurity/state-food-security-nutrition-2018/en/

[76] Ibid.

[77] Ibid.

[78] https://www.cdc.gov/obesity/data/adult.html

[79] http://uconnruddcenter.org/files/Pdfs/TargetedMarketingReport2019.pdf

[80] https://www.cdc.gov/obesity/data/adult.html

[81] Wang YC, McPherson K, Marsh T, Gortmaker SL, Brown M. Health and economic burden of the projected obesity trends in the USA and the UK. Lancet. 2011;378:815-25.

[82] http://www.diabetes.org/advocacy/news-events/cost-of-diabetes.html

[83] http://articles.latimes.com/2003/jun/15/nation/na-diabetes15

[84] https://www.ers.usda.gov/topics/food-nutrition-assistance/food-security-in-the-us/frequency-of-food-insecurity/

[85] https://fns-prod.azureedge.net/sites/default/files/cn/NSLPFactSheet.pdf

[86] https://www.snaptohealth.org/snap/snap-frequently-asked-questions/

[87] https://www.ers.usda.gov/data-products/adoption-of-genetically-engineered-crops-in-the-us/recent-trends-in-ge-adoption.aspx

[88] https://www.cnpp.usda.gov/USDAFoodPlansCostofFood

[89] https://www.ucsusa.org/food_and_agriculture#.XDpBAy2ZPyI

[90] www.med.umich.edu/1libr/Mhealthy/WhatAreEmptyCalories.pdf

[91] http://www.who.int/nutrition/topics/WHO_FAO_ICN2_videos_hiddenhunger/en/

[92] https://www.scientificamerican.com/article/soil-depletion-and-nutrition-loss/

[93] https://www.ncbi.nlm.nih.gov/pmc/articles/PMC4034518/

[94] https://www.cdc.gov/chronicdisease/overview/index.htm

[95] https://news.nationalgeographic.com/news/2005/12/agriculture-food-crops-land/

[96] http://www.fao.org/docrep/014/i2454e/i2454e00.pdf

[97] https://phosphorusalliance.org

[98] Lai, et al, R. Soil Erosion and Land Degradation: The Global Risks. In Soil degradation; Lal, R., Stewart, B.A., Eds.; Springer-Verlag: New York, NY, USA, 1990; pp. 129–172.

[99] World Resources 1996–1997: The Urban Environment; New York, NY, USA, 1997.

[100] Pimentel, D.; Harvey, C.; Resosudarmo, P.; Sinclair, K.; Kurz, D.; McNair, M.; Crist, et al. Environmental and economic costs of soil erosion and conservation benefits. Science **1995**, 267, 1117–1123.

[101] Pimentel, David and Michael Burgess, Soil Erosion Threatens Food Production, *Agriculture*, 2013, 3(3), 443-463.

[102] Eswaran, H.; Lal, R. Land Degradation: An overview. Proceedings of the 2nd International Conference on Land Degradation and Desertification, Khon Kaen, Thailand, 25–29 January 1999; Oxford University Press: New Delhi, India, 2002.

[103] Brady, N. C. *The Nature and Properties of Soils*. New York: Macmillan Publishing Co., 1974.

[104] Gliessman, Stephen R. Agroecology: The Ecology of Sustainable Food Systems, Second Edition, CRC Press; 2 ed., 2006.

[105] Plaster, E. J. *Soil Science and Management*. 3rd ed. Albany: Delmar Publishers, 1996.

[106] https://www.amazon.com/Oceana-Endangered-Oceans-What-Save/dp/B005Q5OYHY https://www.amazon.com/Oceana-Endangered-Oceans-What-Save/dp/B005Q5OYHY

[107] http://www.fao.org/fishery/static/Yearbook/YB2016_USBcard/index.htm

[108] https://www.nationalgeographic.org/projects/pristine-seas/

[109] Unexpected patterns of fisheries collapse in the world's oceans
Malin L. Pinsky, Olaf P. Jensen, Daniel Ricard, Stephen R. Palumbi
Proceedings of the National Academy of Sciences May 2011, 108 (20)

[110] http://www.fao.org/docrep/016/i3010e/i3010e01.pdf

[111] https://www.worldwildlife.org/threats/deforestation

[112] Wye Research and Education Center. Riparian Forest Buffer Panel (Bay Area Regulatory Programs). 2002.

[113] https://www.ipbes.net/news/media-release-worsening-worldwide-land-degradation-now-'critical'-undermining-well-being-32

[114] https://www.saveearth.info/deforestation/

[115] http://www.fao.org/docrep/007/y5609e/y5609e02.htm

[116] https://news.nationalgeographic.com/news/2014/05/140529-conservation-science-animals-species-endangered-extinction/

[117] CD Thomas, et al. Extinction risk from climate change, Nature 427 (6970), 145.

[118] Ibid.

[119] https://www.ers.usda.gov/webdocs/publications/88057/eib-189.pdf?v=0

[120] Losey JE, Vaughan M. The economic value of ecological services provided by insects. Bioscience. 2006;56(4):311–323.

[121] Ibid.

[122] https://www.statista.com/statistics/306105/us-acreage-of-genetically-modified-soybeans/

[123] https://www.nytimes.com/2018/11/05/business/soybeans-farmers-trade-war.html

[124] https://gro-intelligence.com/blog/trump-heads-to-farmers-convention-today?utm_campaign=January%202019%20newsletters&utm2

[125] IARC Monograph, Vol. 112, "Evaluation of 5 organophosphate pesticides and herbicides," Mar 2015.

[126] https://uspirg.org

[127] Myers J P, et al. Concerns over use of glyphosate-based herbicides and risks associated with exposures: a consensus statement. Environ Health, 2016 Feb 17;15:19. doi: 10.1186/s12940-016-0117-0.

[128] Charles Benbrook, "Trends in glyphosate herbicide use in the United States and globally," Environmental Science Europe, Feb. 2, 2016. *at:* https://www.ncbi.nlm.nih.gov/pmc/articles/PMC5044953/

[129] https://uspirg.org

[130] Alexis Temkin, "Breakfast with a dose of Roundup?" Aug. 16, 2018, Environmental Working Group; Ben & Jerry's Statement on Glyphosate, accessed on Nov. 21, 2018.

[131] https://newrepublic.com/article/147066/flints-water-crisis-damage-kids-brains

[132] https://www.ncbi.nlm.nih.gov/pmc/articles/PMC4947579/

[133] http://www.ehsciences.org/

[134] https://www.endocrine.org/

[135] http://www.unep.org/

[136] http://www.alternet.org/environment/female-eggs-found-male-fish-likely-due-pesticide-pollution

[137] https://www.politifact.com/truth-o-meter/statements/2008/aug/22/barack-obama/mccain-has-done-a-little-for-renewable-energy/

[138] https://cropprotectionnetwork.org/2017/04/01/soybean-disease-loss-estimates-from-the-united-states-and-ontario-canada-2015/

[139] https://www.ucdavis.edu/news/genome-sequencing-may-help-avert-banana-armageddon

[140] http://www.fao.org/3/I8656EN/i8656en.pdf

[141] https://www.amazon.com/World-According-Monsanto-Marie-Monique-Robin/dp/1595587098/ref=sr_1_1?ie=UTF8&qid=1503882356&sr=8-1&keywords=monsanto

[142] NIOSH- Agriculture". United States National Institute for Occupational Safety and Health.

[143] Calvert, Geoffrey M. et al.Dec 2008, "Acute pesticide poisoning among agricultural workers in the United States, 1998–2005". American Journal of Industrial Medicine. 51 (12): 883–898.

[144] https://www.epa.gov/nutrientpollution/problem

[145] Walter K. Dodds, et al. Eutrophication of U.S. Freshwaters, Environmental Science & Technology 2009 43 (1), 12-19.

[146] ttp://www.onegreenplanet.org/animalsandnature/facts-on-animal-farming-and-the-environment/

[147] Kansas State University. "Freshwater Pollution Costs US At Least $4.3 Billion A Year." ScienceDaily. ScienceDaily, 17 November 2008. <www.sciencedaily.com/releases/2008/11/081112124418.htm>.

[148] https://www.ewg.org/foodnews/dirty-dozen.php

[149] https://www.foodsafetynews.com/2011/02/fda-confirms-80-percent-of-antibiotics-used-in-animal-ag/

[150] https://www.ucsusa.org/food_and_agriculture/our-failing-food-system/industrial-agriculture/prescription-for-trouble.html#.XA_r2C2ZPyI

[151] https://ycharts.com/indicators/us_corn_acres_planted

[152] https://water.usgs.gov/edu/activity-watercontent.php

[153] https://www.pnas.org/content/110/10/4134

[154] http://edis.ifas.ufl.edu/pi180

[155] https://journals.plos.org/plosone/article?id=10.1371/journal.pone.0185809

[156] https://academic.oup.com/ee/pages/pesticide_exposure_in_non-honey_bees

[157] http://www.pbs.org/wgbh/evolution/library/10/1/l_101_02.html

[158] https://www.canr.msu.edu/grapes/integrated_pest_management/how-pesticide-resistance-develops

[159] Pimentel, D. et al. Water resources: Agricultural and environmental issues. BioScience 2004, 54, 909–918.

[160] https://waterfootprint.org/media/downloads/Hoekstra-Mekonnen-2012-WaterFootprint-of-Humanity.pdf

[161] http://www.waterbenefitshealth.com/water-pollution-facts.html

[162] https://oceanservice.noaa.gov/facts/pollution.html

[163] Ottoman, Michael J. University of Arizona, Arizona Agricultural Extension Service. Personal communication, June, 2007. Verified by Thomas Sands and the irrigation department managers at Salt River Project, Phoenix, AZ, December, 2007.

[164] Hutson, Susan, et. al. "Estimated Use of Water in the U.S. in 2000, " USGA. http://pubs.usgs.gov/circ/2004/circ1268/htdocs/text-ir.html

[165] Orr, John, Turning Crops into Ethanol, (2004), 1. http://www.newfarm.org/features/0804/biofuels/index.shtml

[166] http://www.earth-policy.org/index.php?/plan_b_updates/2001/update1

[167] https://www.ers.usda.gov/data-products/us-bioenergy-statistics/

[168] A Hummer has a 32-gallon tank and gets 10 mpg but requires 13.6 gallons of ethanol per mile.

[169] https://www.scientificamerican.com/article/time-to-rethink-corn/

[170] http://www.fao.org/3/a-i5627e.pdf

[171] //www.nass.usda.gov/Surveys/Guide_to_NASS_Surveys/Chemical_Use/

[172] file:///Users/markedwards/Documents/AMark/Articles B1/SCAD G/Ana Feeds our World/Townsend, Alan R. and Robert W. Howarth. Fixing the Global Nitrogen Problem, Scientific American, Feb 2010, 64-73.

[173] https://www.nap.edu/read/1792/chapter/5#49

[174] https://water.usgs.gov/nawqa/nutrients/pubs/wcp_v39_no12/

[175] http://www.oregon.gov/oha/PH/HealthyEnvironments/DrinkingWater/Monitoring/Documents/health/ammonia.pdf

[176] National Institutes of Health. Air Pollution & Respiratory Disease, July, 2009.

[177] https://news.nationalgeographic.com/2017/12/iowa-agriculture-runoff-water-pollution-environment/

[178] https://www.nrcs.usda.gov/Internet/FSE_DOCUMENTS/stelprdb1041379.pdf

[179] https://www.desmoinesregister.com/story/news/2017/05/25/water-works-plans-15-million-expanded-nitrate-facility/336648001/

[180] https://www.desmoinesregister.com/story/news/2016/08/26/testing-shows-danger-iowas-private-wells/88755480/

[181]http://wwf.panda.org/about_our_earth/teacher_resources/project_ideas/poll ution/

[182] https://transportgeography.org/?page_id=5268

[183] https://ageconomists.com/2018/09/17/us-farm-debt-continues-its-upward-march/

[184]ttps://www.cdc.gov/mmwr/volumes/67/wr/mm6745a1.htm?s_cid=mm6745 a1_w

[185] http://fortune.com/2016/10/27/5-industries-hurt-job/

[186]ttps://www.naturalnews.com/046193_GMO_crop_failures_farmer_suicides _India.html

[187] https://modernfarmer.com/2016/07/farmer-suicide-2/

[188] https://foodprint.org/the-total-footprint-of-our-food-system/issues/sustainable-agriculture/?cid=866

[189] https://www.epa.gov/privatewells

[190] Goyal, SK. A profile of algal biofertilizer. in *Biotechnology of Biofertilizers*, Kannaiyan, S. Ed., Delhi: Narosa Publishing House, 2002, 250 – 258.

[191] www.fao.org/organicag/oa-faq/oa-faq1/en

[192] https://www.resilience.org/resources/resilient-agriculture-cultivating-food-systems-for-a-changing-climate/

[193] https://algae.ucsd.edu/about/why-algae/index.html

[194] https://www.quora.com/How-many-pounds-of-meat-can-you-get-from-a-average-steer

[195] https://www.nytimes.com/2018/08/14/opinion/2020-election-climate-change-trump.html?action=click&pgtype=Homepage&clickSource=story-heading&module=opinion-c-col-right-region&region=opinion-c-col-right-region&WT.nav=opinion-c-col-right-region

[196] https://zeroenergyproject.org/2018/10/07/the-four-zeros/

[197] https://impossiblefoods.com/food/

[198] http://algavia.com

[199] Edwards, Mark. Freedom Foods: Superior Nutrition and Taste from low on the Food Chain for People, Producers and Our Planet, 2011.

[200] Edwards, Mark R. Green Solar Garden: Algae's Promise to End Hunger, 2009.

[201] http://www.aaem.pl/Benefits-and-risks-associated-with-genetically-modified-food-products,71952,0,2.html

[202] Ibid.

[203]ttps://www.sciencedirect.com/science/article/pii/S0013935117300452?via% 3Dihub

[204] Center for Food Safety, http://www.centerforfoodsafety.org /2011/03/18/

[205] N. Mayot, P. Matrai, I. et al. Assessing Phytoplankton Activities in the Seasonal Ice Zone of the Greenland Sea Over an Annual Cycle. *Journal of Geophysical Research: Oceans*, 2018; 123 (11)

[206] WWAP (2017). The United Nations World Water Development Report 2017. Wastewater: The Untapped Resource. Paris: United Nations World Water Assessment Programme, UNESCO.

[207] https://books.google.com/books?hl=en&lr=&id=QSxWUoj9twQC&oi=fnd &pg=PR7&dq=social+justice+food&ots=HKP6vLv5rj&sig=FbDtXN0YE DwDT0LDecHoXm4CoRc#v=onepage&q=social%20justice%20food&f=fa lse

[208] http://nytimes.com/.../global-food-prices-on-the-rise-united-nations

[209] Warner, Jennifer. CDC: Kids Lack Access to Healthy Food Choices, WebMD Health News, April 26, 2011.

[210] Landers TF, et al. A review of antibiotic use in food animals: perspective, policy, and potential. *Public Health Rep.* 2012;127(1):4-22.

[211] https://www.reuters.com/article/us-health-superbugs-idUSKBN161107

[212] http://www.cidrap.umn.edu/news-perspective/2017/02/multidrug-resistant-infections-rising-us-kids

[213] https://www.ncbi.nlm.nih.gov/pmc/articles/PMC4144270/

[214] https://research.jcu.edu.au/.../copy2_of_Bioremediationforcoalfiredpowerst ations.pdf

[215] https://www.ncbi.nlm.nih.gov/pmc/articles/PMC4052567/

[216] https://www.sciencedirect.com/science/article/pii/S2314853515000621

[217] esciencecentral.org/ebooks/recent-advances-in-microalgal-biotechnology/adsorption-and-absorption-of-heavy-metals.php

[218] www.mdpi.com/1996-1073/9/3/132/pdf

[219] www.arb.ca.gov/.../carbon_capture_corp.pdf

[220] Correia, Andrew et at. The Effect of Air Pollution Control on Life Expectancy in the United States, Epidemiology, December 3, 2012.

[221] Fann et al., 2012, Estimating the National Public Health Burden to Ambient $PM_{2.5}$ and Ozone, Risk Analysis 32(1) 81-95.

[222] https://www.nal.usda.gov/afsic/organic-production

[223] https://www.ers.usda.gov/data-products/organic-production/

[224] https://regenerationinternational.org/why-regenerative-agriculture/

[225] https://regenerationinternational.org/2015/10/06/hungry-for-land-small-farmers-feed-the-world-with-less-than-a-quarter-of-all-farmland/

226 https://regenerationinternational.org/2018/10/09/reversing-climate-change-through-regenerative-agriculture/
227 https://www.noble.org/news/publications/ag-news-and-views/2011/april/have-you-seen-the-price-of-nitrogen-fertilizer/
228 Edwards, Mark R. Sustainable Functional Foods and Medicines support Vitality, Sex, and Longevity for a 100-year Starship Expedition, *Journal of the British Interplanetary Society*, June 2013.
229 http://vsemart.com/rare-art-diatom-arrangement-klaus-kemp/
230 http://tmagazine.blogs.nytimes.com/2014/09/16/diatomist-film-matthew-killip-premiere/?_r=0
231 https://www.khanacademy.org/science/biology/bacteria-archaea/prokaryote-structure/v/prokaryotic-and-eukaryotic-cells
232 https://www.livescience.com/55008-lichens.html
233 http://msnucleus.org/watersheds/biological/microalgaegen.html
234 https://utex.org/pages/algal-collections
235 https://www.khanacademy.org/test-prep/mcat/cells/eukaryotic.../a/organelles-article
236 https://www.nature.com/scitable/topicpage/cell-membranes-14052567
237 UTEX Algae Culture website. http://www.utex.org/
238 Graham, L.E. The origin of the life cycle of land plants. *American Scientist*, 1985, 73; 78 – 96.
239 David Hall et al. Institute of Biology (1999). Photosynthesis. Cambridge.
240 Ibid.
241 Pimentel, David, Marcia Pimentel, Food, Energy and Society, 1979, 37.
242 https://www.eatthismuch.com/food/nutrition/ear-of-corn,609100/
243 https://pubs.acs.org/doi/pdf/10.1021/ac202668c
244 https://www.ncbi.nlm.nih.gov/books/NBK26869/
245 https://www.ncbi.nlm.nih.gov/pubmed/24441762
35 https://www.sciencedirect.com/topics/medicine-and-dentistry/mutagenesis-molecular-biology-technique
247 http://www.soes.soton.ac.uk/staff/tt/
248 https://phys.org/news/2014-09-algae-oceans.html
249 https://www.amazon.com/Freedom-Foods-Superior-People-Producers/dp/1461158583
250 https://www.ncbi.nlm.nih.gov/pubmed/18996876
251 https://www.organic-center.org/reportfiles/Davis_ppt.pdf
252 https://www.theguardian.com/lifeandstyle/2015/feb/10/nutrition-hunger-food-children-vitamins-us

[253] http://www.algaeindustrymagazine.com/algae-solutions-for-pesticide-exposure-and-autism/

[254] https://www.theguardian.com/lifeandstyle/nutrition

[255] https://www.organic-center.org/reportfiles/Davis_ppt.pdf

[256] http://www.npr.org/sections/thesalt/2015/08/11/429985941/protein-goes-green-can-algae-become-the-next-soy

[257] https://www.organic-center.org/reportfiles/Davis_ppt.pdf

[258] ttp://www.fao.org/fileadmin/templates/nr/sustainability_pathways/docs/Factsheet_SMALLHOLDERS.pdf

[259] https://www.ncbi.nlm.nih.gov/pubmed/24261532

[260] https://www.researchgate.net/publication/223873726_A_physiological_approach_for_preparing_and_conducting_intestinal_bioavailability_studies_using_experimental_system

[261] https://www.webmd.com/food-recipes/news/20120106/antibiotics-food-animals-faq#1

[262] M. Lamminen, et al. Different microalgae species as a substitutive protein feed for soya bean meal in dairy cow diets, Animal Feed Science and Technology, 247, 2019, 112-126.,

[263] Emma M Brown, et al. Seaweed and human health, *Nutrition Reviews*, 72:3, March, 2014, 205–216

[264] https://www.ncbi.nlm.nih.gov/pubmed/19695833

[265] https://onlinelibrary.wiley.com/doi/pdf/10.1002/elsc.201400191

[266] http://onlinelibrary.wiley.com/wol1/doi/10.1002/elsc.201400191/full

[267] https://onlinelibrary.wiley.com/doi/pdf/10.1002/elsc.201400191

[268] https://onlinelibrary.wiley.com/doi/pdf/10.1002/elsc.201400191

[269] https://www.ncbi.nlm.nih.gov/pmc/articles/PMC3667473/

[270] https://www.pnas.org/content/115/49/12401

[271] https://www.amazon.com/Freedom-Foods-Superior-Producers-Strategy-ebook/dp/B009WSHX42/ref=sr_1_1?ie=UTF8&qid=1525045688&sr=8-1&keywords=freedom+foods+edwards&dpID=51yFYM7FBML&preST=_SY445_QL70_&dpSrc=srch

[272] https://www.marchofdimes.org/mission/prematurity-campaign.aspx

[273] https://www.ncbi.nlm.nih.gov/pmc/articles/PMC5387034/

[274] https://www.sciencedirect.com/topics/neuroscience/phytic-acid

[275] https://academic.oup.com/jn/article/137/12/2691/4670055

[276] https://www.researchgate.net/publication/279694812_Marine_algae_as_natural_source_of_iodine_in_the_feeding_of_freshwater_fish_-_A_new_possibility_to_improve_iodine_supply_of_man

[277] https://www.researchgate.net/publication/12664301_A_taste_for_umami

[278] http://onlinelibrary.wiley.com/doi/10.1002/elsc.201400191

[279] https://www.journals.elsevier.com/journal-of-functional-foods/

[280] http://www.ashwagandhaadvantage.com/reports/2017/06/enhancing-nutrition.aspx?cmpid=FUNCTIONRT_hyper7-18

[281] http://www.xtcworldinnovation.com/default.asp?id=21

[282] https://www.elsevier.com/...algae-for-foods-and-nutraceuticals/.../978-0-85709-512-1

[283] www.foodnavigator-usa.com/feature/news-by-month/08/2016)

[284] www.foodnavigator-usa.com/feature/news-by-month/08/2016)

[285] forbes.com/sites/deeppatel/2017/08/26/food-leaders-take-notice-how-millennials-are-changing-the-way-we-eat/

[286] http://www.ift.org/newsroom/news-releases/2018/april/21/top-ten-functional-food-trends-for-2014.aspx

[287] https://www.tandfonline.com/doi/abs/10.1080/10447318.2018.1451429

[288] Ibid.

[289] https://www.sciencedirect.com/science/article/pii/S0002934301009950

[290] http://pubs.sciepub.com/jfnr/3/5/9/

[291] https://link.springer.com/chapter/10.1007/978-94-009-2049-1_14

[292] https://www.sciencedirect.com/science/article/pii/B9780123876690000193

[293] https://www.ncbi.nlm.nih.gov/pmc/articles/PMC5744479/

[294] https://www.ncbi.nlm.nih.gov/pmc/articles/PMC4158787/

[295] http://onlinelibrary.wiley.com/doi/10.1002/9780470277676.fmatter/pdf

[296] https://www.researchgate.net/profile/Wim_Verbeke2/publication/2235241
36_Functional_Foods_Consumer_Willingness_to_Compromise_on_Taste
_for_Health/links/02e7e530caf304b1e9000000.pdf

[297] https://www.ncbi.nlm.nih.gov/pmc/articles/PMC5387034/

[298] http://www.scirp.org/(S(vtj3fa45qm1ean45vvffcz55))/reference/References
Papers.aspx?ReferenceID=1125722

[299] https://pdfs.semanticscholar.org/636c/21676ce49ea929b1a1393d6110009d
9fe44a.pdf

[300] http://(www.huffingtonpost.com/2014/07/24/seaweed-beer_n_5614794.html).

[301] https://www.researchgate.net/profile/Laurie_Eve_Rioux

[302] http://www.npr.org/2012/10/16/163002343/test-kitchen-chefs-talk-the-science-of-savory

[303] https://eatforum.org/eat-lancet-commission/

[304] https://www.ncbi.nlm.nih.gov/pmc/articles/PMC3840875/

[305] Donovan A. Argueta, Nicholas V. DiPatrizio. Peripheral endocannabinoid signaling controls hyperphagia in western diet-induced obesity. *Physiology & Behavior*, 2017; 171: 32

[306] https://www.ncbi.nlm.nih.gov/pmc/articles/PMC4034518/

[307] http://www.algaeindustrymagazine.com/department/algae-secrets/

[308] Black RE, et al. *Lancet.* Maternal and child undernutrition and overweight in low-income countries. 2013 Aug 3;382(9890):427-451.

[309] https://thousanddays.org/the-issue/stunting/

[310] Richard SA, Black RE, Gilman RH, et al. Wasting is associated with stunting in early childhood. *J Nutr.* 2012;142(7):1291-6.

[311] http://www.cdc.gov/reproductivehealth/maternalinfanthealth/PretermBirth.htm

[312] https://www.cdc.gov/reproductivehealth/maternalinfanthealth/pretermbirth.htm

[313] Boyle CA, et al. Trends in the Prevalence of Developmental Disabilities in US Children, 1997–2008. Pediatrics. 2011; 27: 1034-1042.

[314] J. A. Grieger, et al. **Preconception Dietary** Patterns in Human Pregnancies Are Associated with Preterm Delivery. *Journal of Nutrition*, 2014.

[315] https://www.ncbi.nlm.nih.gov/pubmed/1570808

[316] https://journals.sagepub.com/doi/abs/10.1177/1362361302006001008

[317] Rebecca J. Schmidt, et al. Maternal Intake of Supplemental Iron and Risk of Autism Spectrum Disorder, American Journal of Epidemiology, 180: 9, 1 November 2014, 890–900.

[318] http://www.massgeneral.org/children/services/lurie-center/Omega-3.aspx

[319] http://www.gmhcn.org/files/Wellness/Omega-3forDepressionandBipolarDisorder.html

[320] http://irp.nih.gov/pi/joseph-hibbeln

[321] http://www.nih.gov/news/health/aug2011/niaaa-23.htm

[322] http://www.gmhcn.org/files/Wellness/Omega-3forDepressionandBipolarDisorder.html

[323] https://www.ncbi.nlm.nih.gov/pmc/articles/PMC2661342/

[324] Birth Defects Research (Part A): Clinical and Molecular Teratology 2010, 88, 1008-16.

[325] https://www.ncbi.nlm.nih.gov/pmc/articles/PMC4046315/

[326] Romero R, Gotsch F, Pineles B, Kusanovic JP. Inflammation in pregnancy: its roles in reproductive physiology, obstetrical complications, and fetal injury. Nutr Rev. 2007;65(12 pt 2):S194–S202 [PubMed]

[327] http://www.aafa.org/page/allergy-facts.aspx

[328] Yoshihiro Miyake, et al. Dietary Intake of Seaweed and Minerals and Prevalence of Allergic Rhinitis in Japanese Pregnant Females: Baseline Data From the Osaka Maternal and Child Health Study, Annals of Epidemiology, 16:8, August 2006, 614-621.

[329] //www.ncbi.nlm.nih.gov/pmc/articles/PMC2190961/pdf/je17741199.pdf

[330] http://nopr.niscair.res.in/bitstream/123456789/41113/1/IJMS%2046%284%29%20663-667.pdf

[331] https://www.researchgate.net/publication/225286713_Biological_activities_and_potential_health_benefits_bioactive_peptides_derived_from_marine_organisms

[332] Gregory, J.A., Mayfield, et al. Immunotherapy using algal-produced Ara h 1 core domain suppresses peanut allergy in mice. Plant Biotechnology Journal, 14677644, Jul 2016, Vol. 14:7.

[333] https://books.google.com/books?isbn=0123877008

[334] https://www.ncbi.nlm.nih.gov/pubmed/26308008

[335] Quang-To Le, et al. Inhibitory effects of polyphenols isolated from marine alga Ecklonia cava on histamine release, Process Biochemistry, 44:2, February 2009, 168-176.

[336] https://www.cdc.gov/diabetes/basics/diabetes.html

[337] http://www.diabetes.org/advocacy/news-events/cost-of-diabetes.html

[338] https://pharm.ucsf.edu/kidney/need/statistics

[339] https://www.sciencedirect.com/science/article/pii/S0367326X13000567

[340] https://link.springer.com/article/10.1007/s10811-010-9556-0

[341] Seidell JC, Halberstadt J. The global burden of obesity and the challenges of prevention. Ann Nutr Metab. 2015;66(suppl 2):7–12.

[342] https://www.ncbi.nlm.nih.gov/pmc/articles/PMC3517813/

[343] //norml.org/library/item/introduction-to-the-endocannabinoid-system

[344] http://www.drugdevelopment-technology.com/projects/rimonabant/

[345] https://www.ncbi.nlm.nih.gov/pubmed/19578688

[346] http://www.drugdevelopment-technology.com/projects/rimonabant/

[347] https://www.nature.com/articles/ijo2008235

[348] http://www.diabetesincontrol.com/alginate-gels-sea-weed-may-reduce-blood-sugar-spikes/

[349] shura.shu.ac.uk/1182/1/alginate_reduces_the_increased_uptake_for_shura.pdf

[350] https://www.ncbi.nlm.nih.gov/pubmed/9119615

[351] https://www.ncbi.nlm.nih.gov/pubmed/28414549

[352] https://nccih.nih.gov/health/omega3/introduction.htm

[353] https://www.ncbi.nlm.nih.gov/pubmed/19335713

[354] http://health.usnews.com/health-conditions/heart-health/lowering-ldl-cholesterol

[355] http://www.pfigueiredo.org/TA23.pdf

[356] https://www.sciencedirect.com/science/article/pii/S1876610217303247

[357] //scholarsarchive.byu.edu/cgi/viewcontent.cgi?article=4141&context=etd

[358] http://www.algaeindustrymagazine.com/can-algae-save-children-from-heavy-metals-poisoning/

[359] http://www.toxicsaction.org/problems-and-solutions/pesticides

[360] http://www.algaeindustrymagazine.com/algae-solutions-for-pesticide-exposure-and-autism/

[361] https://www.ncbi.nlm.nih.gov/pmc/articles/PMC5555614/

[362] https://www.ncbi.nlm.nih.gov/pmc/articles/PMC2267278/

[363] ttps://www.umms.org/ummc/health/medical/altmed/supplement/omega6-fatty-acids

[364] http://www.sciencedirect.com/science/article/pii/S1323893014000100

[365] https://www.researchgate.net/publication/8450096_Ratio_of_omega-6_to_omega-3_fatty_acids_and_childhood_asthma

[366] http://archive.inside.indiana.edu/editors-picks/health-wellness/2013-05-30-iniub-health-omega-3.shtml

[367] https://journals.lww.com/acsm-ealthfitness/fulltext/2012/09000/Fish_Oil_Supplementation_and_Athletic_Performance.9.aspx

[368] https://journals.lww.com/acsm-healthfitness/fulltext/2012/09000/Fish_Oil_Supplementation_and_Athletic_Performance.9.aspx

[369] http://www.sciencedirect.com/science/article/pii/S1525861013003265

[370] www.mdpi.com/2304-8158/6/5/33/htm

[371] https://health.clevelandclinic.org/2016/09/how-omega-3-foods-can-help-you-be-a-better-athlete/

[372] https://www.ncbi.nlm.nih.gov/pmc/articles/PMC3576896/

[373] https://www.liebertpub.com/doi/abs/10.1089/152308603321223630

[374] https://www.fda.gov/food/ingredientspackaginglabeling/gras/

[375] https://www.ncbi.nlm.nih.gov/pmc/articles/PMC5447909/

[376] https://www.ncbi.nlm.nih.gov/pmc/articles/PMC5537823/

[377] https://www.infona.pl/resource/bwmeta1.element.elsevier-752b1a9f-3f2f-3e7b-915f-f2f7e4c548b7

[378] gepea.fr/files/Rapport-d-activite-2012-2013-3-fr-4.pdf

[379] http://algavia.com/

[380] https://www.ncbi.nlm.nih.gov/pmc/articles/PMC4425044/

[381] https://blogs.scientificamerican.com/artful-amoeba/in-a-world-of-shifting-sand-algae-turn-to-dark-fermentation/

[382] Rheinheimer G (1992) Aquatic microbiology, 3rd edn. Wiley, New YorkGoogle Scholar

[383] Lane A, Kubanek J et al. (2009) Desorption electrospray ionization mass spectrometry reveals surface-mediated antifungal chemical defense of a tropical seaweed. PNAS 106:7314–7319.

[384] https://www.ncbi.nlm.nih.gov/pmc/articles/PMC3027850/

[385] https://www.ncbi.nlm.nih.gov/pmc/articles/PMC4782487/

[386] Mikinori Ueno et al. Acidic polysaccharides isolated from marine algae inhibit the early step of viral infection, International Journal of Biological Macromolecules, 124:1, March 2019, 282-290.

[387] Thuy, Thanh, et al. Anti-**HIV** activity of fucoidans from three brown seaweed species. Carbohydrate Polymers, Jan 2015, 115, 122-128.

[388] Teas, J. & Irhimeh, M.R. Dietary algae and HIV/AIDS: proof of concept clinical data, Journal of Applied Phycology, J Appl Phycol (June, 2012) 24:3, 575–582.

[389] UNAIDS (2011) UNAIDS Joint United Nations Programme on HIV/AIDS. http://www.unaids.org/en/dataanalysis/epidemiology/2009aids epidemicupdate/

[390] https://www.cdc.gov/nchs/fastats/obesity-overweight.htm

[391] https://onlinelibrary.wiley.com/doi/pdf/10.1002/elsc.201400191

[392] https://www.ncbi.nlm.nih.gov/pubmed/19375880

[393] apjcn.nhri.org.tw/server/apjcn/18/2/index.php

[394] https://www.triphobo.com/blog/best-molecular-gastronomy-restaurants

[395] http://www.molecularrecipes.com/molecular-gastronomy-restaurants-molecular-mixology-bars/united-states/

[396] http://www.gayot.com/restaurants/best-us-top10-molecular-gastronomy_01us.html

[397] http://www.reuters.com/article/us-seaweed-weightloss-idUSBRE85C1DD20120613

[398] ashlandmarinebio.blogspot.com/2010/

[399] https://www.ncbi.nlm.nih.gov/pmc/articles/PMC3654245/

[400] http://naturalsociety.com/5-grams-spirulina-daily-reverse-radiation-damage/

[401] https://www.ncbi.nlm.nih.gov/pubmed/15163472

[402] https://www.ncbi.nlm.nih.gov/pubmed/17390632

[403] https://www.cdc.gov/bloodpressure/index.htm

[404] http://www.clevelandheartlab.com/blog/horizons-omega-3-fatty-acids-a-natural-way-to-lower-blood-pressure/

[405] https://ods.od.nih.gov/factsheets/Omega3FattyAcids-HealthProfessional/

[406] https://www.ncbi.nlm.nih.gov/pmc/articles/PMC3257651/

[407] https://www.ncbi.nlm.nih.gov/pubmed/28779377

[408] Personal observation.

[409] scielo.isciii.es/scielo.php?script=sci_arttext&pid=S0212-16112012001100005

[410] http://onlinelibrary.wiley.com/doi/10.1002/elsc.201400191/abstract

[411] pubs.acs.org/doi/abs/10.1021/jf901070g

[412] https://www.ncbi.nlm.nih.gov/pubmed/15702613

[413] https://www.sciencedaily.com/releases/2012/07/120706194954.htm

[414] Ibid.

[415] https://www.ncbi.nlm.nih.gov/pmc/articles/PMC5516777/

[416] https://www.nhs.uk/conditions/anticoagulants/

[417] https://www.ncbi.nlm.nih.gov/pmc/articles/PMC3278069/

[418] http://www.strokecenter.org/patients/about-stroke/ischemic-stroke/

[419] https://www.researchgate.net/profile/Isuru_Wijesekara/publication/251510268_Review_Biological_activities_and_potential_health_benefits_of_sulfated_polysaccharides_derived_from_marine_algae/links/560be47108aea68653d2cbc0/Review-Biological-activit

[420] https://www.ncbi.nlm.nih.gov/pmc/articles/PMC3014083/

[421] https://pdfs.semanticscholar.org/8b54/e3ca3c2b215b7e0a4fd83a4dc1a292e1c86c.pdf

[422] https://www.ncbi.nlm.nih.gov/pubmed/22054951

[423] https://www.ncbi.nlm.nih.gov/pubmed/23815295

[424] https://www.ncbi.nlm.nih.gov/pmc/articles/PMC4736713/

[425] https://www.ncbi.nlm.nih.gov/pmc/articles/PMC4532706/

[426] https://www.ncbi.nlm.nih.gov/pmc/articles/PMC4532706/

[427] https://www.ncbi.nlm.nih.gov/pmc/articles/PMC5490612/#B6-nutrients-09-00633

[428] http://www.mdpi.com/2072-6643/9/6/633/htm

[429] https://jmedicalcasereports.biomedcentral.com/articles/10.1186/1752-1947-4-103

[430] https://uclouvain.be/en/research-institutes/ldri/about.html

[431] https://www.ncbi.nlm.nih.gov/pubmed/?term=Neyrinck%20AM%5BAuthor%5D&cauthor=true&cauthor_uid=28632181

[432] https://www.ncbi.nlm.nih.gov/pmc/articles/PMC5490612/

[433] https://www.cancer.org/research/cancer-facts-statistics.html

[434] https://www.ncbi.nlm.nih.gov/pmc/articles/PMC3901206/

[435] www.vjs.ac.vn/index.php/vjbio/article/view/6630

[436] http://www.independent.co.uk/news/science/genetically-engineered-algae-cancer-treatment-chemotherapy-a6732226.html

[437] https://ir.nctu.edu.tw/bitstream/11536/7062/1/000265810900036.pdf

[438] https://www.ncbi.nlm.nih.gov/pmc/articles/PMC3210606/

[439] https://www.ncbi.nlm.nih.gov/pubmed/?term=Suh%20SS%5BAuthor%5D&cauthor=true&cauthor_uid=28638272

[440] https://www.ncbi.nlm.nih.gov/pmc/articles/PMC5479125/

[441] https://www.mdpi.com/1420-3049/20/4/7097/pdf

[442] https://www.researchgate.net/publication/11307776_Bioactive_steroids_from_the_brown_Alga_Sargassum_carpophyllum

[443] www.mdpi.com/2072-6651/4/11/1058

[444] https://www.ncbi.nlm.nih.gov/pmc/articles/PMC3538218/

[445] http://www.pnas.org/content/110/1/E15.full

[446] http://ucsdnews.ucsd.edu/pressrelease/biologists_engineer_algae_to_make_complex_anti_cancer_designer_drug

[447] www.pnas.org/content/110/1/E15/tab-article-info

[448] www.latimes.com/.../la-ss-ucsd-how-algae-could-change-the-world-dto-story.html

[449] http://ucsdnews.ucsd.edu/pressrelease/biologists_engineer_algae_to_make_complex_anti_cancer_designer_drug

[450] https://www.ncbi.nlm.nih.gov/pubmed/26137911

[451] http://www.physorg.com/tags/blindness/

[452] http://www.ncbi.nlm.nih.gov/pubmedhealth/PMH0002024/

[453] www.blindness.org/retinitis-pigmentosa

[454] releases.jhu.edu/2010/07/14/blind-mice-can-see-thanks-to-special-retinal-cells/

[455] http://www.usc.edu/schools/medicine/research/institutes/igm/

[456] https://www.technologyreview.com/s/421400/brain-control/

[457] https://ntrs.nasa.gov/archive/nasa/casi.ntrs.nasa.gov/19830012262.pdf

[458] https://www.ncbi.nlm.nih.gov/pubmed/27682165

[459] http://valensa.com/

[460] http://lpi.oregonstate.edu/infocenter/glossary.html#meta-analysis

[461] https://bmcmusculoskeletdisord.biomedcentral.com/articles/10.../1471-2474-11-136

[462] http://www.umm.edu/altmed/articles/omega-3-000316.htm#ixzz2KPrT4x7M

463 https://www.ncbi.nlm.nih.gov/pmc/articles/PMC4557028/

464 http://pubs.acs.org/doi/abs/10.1021/jf201114d

465 http://pubs.acs.org/doi/abs/10.1021/jf502420h

466 http://onlinelibrary.wiley.com/doi/10.1002/elsc.201400191/abstract

467 https://www.ncbi.nlm.nih.gov/pmc/articles/PMC3210606/

468 https://www.ncbi.nlm.nih.gov/pubmed/28699508

469 https://www.ncbi.nlm.nih.gov/pmc/articles/PMC3925837/

470 https://www.researchgate.net/publication/259699730_Developing_inexpen sive_malaria_vaccines_from_plants_and_algae

471 https://www.sciencedaily.com/releases/2012/05/120516174437.htm

472 Ibid.

473 https://health.ucsd.edu/news/releases/Pages/2016-04-07-human-vaccines-project-and-mesa-consortium.aspx

474 https://www.aphis.usda.gov/plant_health/plant_pest_info/weeds/.../PPATe xt.pdf

475 https://btiscience.org/

476 https://www.ncbi.nlm.nih.gov/pmc/articles/PMC5445130/

477 https://link.springer.com/article/10.1007%2Fs00253-011-3279-2

478 https://www.ncbi.nlm.nih.gov/pubmed/24659086

479 https://www.broadinstitute.org/what-broad/areas-focus/project-spotlight/questions-and-answers-about-crispr

480 https://www.nature.com/cr/journal/vaop/ncurrent/full/cr201757a.html

481 http://www.sciencedirect.com/science/article/pii/S2211926417300024

482 https://www.who.int/mediacentre/news/releases/2014/air-pollution/en/

483 http://documents.worldbank.org/curated/en/781521473177013155/pdf/10 8141-REVISED-Cost-of-PollutionWebCORRECTEDfile.pdf

484 FabioCaiazzo, Air pollution and early deaths in the United States. Part I: Quantifying the impact of major sectors in 2005, Atmospheric Environment, 79, November 2013, 198-208.

485 https://www.who.int/gho/phe/air_pollution_pm25_concentrations/en/

486 http://www.nature.com/nature/journal/v525/n7569/full/nature15371.html

487 http://www.nature.com/nature/journal/v525/n7569/full/nature15371.html

488 https://www.ucsusa.org/clean-energy/coal-and-other-fossil-fuels/coal-air-pollution#.XEcEyy2ZPx4

489 http://www.sciencedirect.com.ezproxy1.lib.asu.edu/science/article/pii/S092 5857416305778

490 Ibid.

491 https://phys.org/news/2013-03-algae-capture-co2.html#jCp

[492] https://www.physicsforums.com/threads/algae-producing-oxygen-in-diving-cylinder.636257/

[493] https://www.netl.doe.gov/project-information?k=FE0026490&show=ppp

[494] Colin M. Beal, Ian Archibald, Mark E. Huntley, Charles H. Greene, Zackary I. Johnson. Integrating Algae with Bioenergy Carbon Capture and Storage (ABECCS) Increases Sustainability. *Earth's Future*, 2018

[495] http://www.powermag.com/breakthrough-carbon-capturing-algae-project/

[496] http://www.globalgae.com/news

[497] http://www.powerplantccs.com/ccs/cap/fut/alg/alg_ccs_history.html

[498] https://energyfactor.exxonmobil.com/science-technology/fuel-cell-energy/?utm_source=bing&utm_medium=cpc&utm_campaign=XOM+%7C+Corp+%7C+ELH+%7C+Traffic+%7C+Non+Brand+%7C+Technology+%7C+Carbon+Capture+%7C+BMM&utm_content=Non+Brand+%7C+Carbon+%7C+Capture&utm_term=%2Bcarbon+%2Bcapture&gclid=CNjEkozbgeACFYSExQId9sgMZw&gclsrc=ds

[499] https://www.cell.com/joule/fulltext/S2542-4351(18)30225-3

[500] Elena Villar-Navarro, et al. Removal of pharmaceuticals in urban wastewater, Water Research, 139, 1 August 2018, 19-29.

[501] https://nioo.knaw.nl/en

[502] Closing Domestic Nutrient Cycles Using Microalgae, Tânia Vasconcelos Fernandes, et al. Environ. Sci. Technol., 2015, 49 (20), pp 12450–12456, DOI: 10.1021/acs.est.5b02858

[503] VO HOANG NHAT, P. et al. Can algae-based technologies be an affordable green process for biofuel production and wastewater remediation? **Bioresource Technology**, [s. l.], v. 256, p. 491–501, 2018.

[504] https://www.worldwildlife.org/stories/the-impact-of-a-cotton-t-shirt

[505] Rita Kant, Textile dyeing industry an environmental hazard, 4:1, 22-26 (2012) Natural Science http://dx.doi.org/10.4236/ns.2012.41004

[506] HOLKAR, C. R. et al. A critical review on textile wastewater treatments: Possible approaches. Journal of Environmental Management. Nov 2016, Vol. 182, p351-366

[507] https://worldwatertechinnovation.com/speakers/frank-rogalla-2/

[508] https://www.amazon.com/Smartcultures-Environmental-Pollution-Regenerates-Ecosystems/dp/1456524690

[509] https://oceanservice.noaa.gov/facts/phyto.html

[510] www.ucmp.berkeley.edu/bacteria/cyanointro.html

[511] https://www.nature.com/scitable/knowledge/.../biological-nitrogen-fixation-23570419

[512] https://www.smithsonianmag.com/smart-news/see-how-ladybugs-fit-wings-inside-their-spotted-red-shells-180956366/

[513] https://www.agcensus.usda.gov/Newsroom/2015/08_31_2015.php

[514] Kolbert, Elizabeth. *Field Notes from a Catastrophe: Man Nature and Climate Change*. New York: Bloomsbury, 2006: 96-97.

[515] https://www.pnas.org/content/early/2019/01/08/1812883116

[516] Eric Rignot, et al. Four decades of Antarctic Ice Sheet, Proceedings of the National Academy of Sciences Jan 2019, 116 (4) 1095-1103.

[517] https://www.worldwildlife.org/threats/soil-erosion-and-degradation

[518] Soil as an Endangered Ecosystem", David Pimental, Bioscience; Nov 2000

[519] //www.iowalearningfarms.org/files/page/files/Cost_of_Eroded_Soil.pdf

[520] World Watch. *Cropland Losses Threaten World Food Supplies*, July 27, 1996. http://www.worldwatch.org/cropland-losses/

[521] Edwards, Mark R. *Smartcultures: Sustainable Food despite Climate Change*, 2010.

[522] //www.scientificamerican.com/article/the-internet-of-things-goes-nano/

[523] https://www.forbes.com/sites/matthewherper/2012/02/10/the-truly-staggering-cost-of-inventing-new-drugs/#4e0730864a94

[524] https://www.fitbit.com/

[525] https://www.wired.com/insights/2014/11/the-internet-of-things-bigger/

[526] http://www.nowtheendbegins.com/wisconsin-company-three-square-market-becomes-first-united-states-microchip-employees/

[527] https://www.fda.gov/aboutfda/transparency/basics/ucm192695.htm

[528] http://www.goedomega3.com/about-us

[529] https://www.youtube.com/watch?v=KTAg936uoic&

[530] http://www.sciencedirect.com/science/article/pii/0304416584901211

[531] http://www.precisionnutrition.com/eating-bugs

[532] http://onlinelibrary.wiley.com/doi/10.1002/elsc.201400191/pdf

[533] https://www.amazon.com/Biowar-I-Hunger-Mark-Edwards/dp/1435700198

[534] https://www.bloomberg.com/news/articles/2017-06-22/trump-pledges-support-for-under-siege-ethanol-amid-epa-delays

[535] Ibid.

[536] http://www.algaeindustrymagazine.com/algae-sex-part-4/

[537] https://www.fastcompany.com/1680051/all-the-water-it-takes-to-produce-a-burger

[538] **https://www.youtube.com/watch?v=7jrhCMBNz1o&feature=youtu.be**

[539] https://www.gp.org/green_new_deal

[540] https://www.who.int/en/news-room/fact-sheets/detail/drinking-water

[541] https://www.thelancet.com/journals/laninf/article/PIIS1473-3099(15)00349-7/fulltext

[542] https://www.cawrecycles.org/dairy-waste-pollution

[543] https://www.thelancet.com/journals/lancet/article/PIIS0140-6736(18)31788-4/fulltext

[544] https://www.ncbi.nlm.nih.gov/pmc/articles/PMC4034518/

[545] https://www.ncbi.nlm.nih.gov/pmc/articles/PMC4032627/

[546] https://www.fda.gov/food/resourcesforyou/consumers/ucm079311.htm

[547] http://farrp.unl.edu/informallbig8

[548] https://www.fda.gov/food/resourcesforyou/consumers/ucm079311.htm

[549] http://www.huffingtonpost.com/entry/oh-lovely-the-tick-that-gives-people-meat-allergies_us_59496743e4b0579a1f392748

[550] https://oceanservice.noaa.gov/facts/redtide.html

[551] http://www.whoi.edu/redtide/

[552] http://www.npr.org/sections/thesalt/2013/10/11/232106472/what-s-in-that-chicken-nugget-you-really-don-t-want-to-know

[553] globalnutritionreport.org

[554] https://newfoodeconomy.org/agriculture-industry-influence-money-academic-research/

[555] https://newfoodeconomy.org/agriculture-industry-influence-money-academic-research/

[556] https://newfoodeconomy.org/author/kate-cox/

[557] https://eatforum.org/eat-lancet-commission/

[558] https://www.cdc.gov/healthcommunication/toolstemplates/entertainmented/tips/fooddesert.html

[559] https://www.cdc.gov/nchs/fastats/health-expenditures.htm

[560] http://www.huffingtonpost.com/2011/09/22/farm-subsidies-junk-food_n_975711.html

[561] U.S. Department of Agriculture, "Farm Household Income and Characteristics," www.ers.usda.gov/data-products/farm-household-income-and-characteristics.aspx. See spreadsheet for "Mean and Median Farm Operator Household Income and Ratio of Farm Household to U.S. Household Income, 1960–2016.

[562] U.S. Department of Agriculture, "2012 Census Highlights: Farm Economics," ACH12-2, May 2014. See also Environmental Working Group, Farm Subsidy Database, https://farm.ewg.org.

563 Vincent H. Smith, Joseph W. Glauber, Barry K. Goodwin, and Daniel A. Sumner, "Agricultural Policy in Disarray: Reforming the Farm Bill — An Overview," American Enterprise Institute, October 2017, p. 20.

564 Ibid.

565 https://www.ewg.org

566 https://usda.library.cornell.edu

567 https://www.sacbee.com/news/state/california/water-and-drought/article103987631.html

568 https://www.ppic.org/publication/water-and-the-future-of-the-san-joaquin-valley/

569 https://www.ppic.org/content/pubs/report/R_314EHR.pdf

570 https://www.nytimes.com/2016/08/06/upshot/do-oil-companies-really-need-4-billion-per-year-of-taxpayers-money.html

571 https://www.taxpayer.net/energy-natural-resources/subsidy-gusher-taxpayers-stuck-with-massive-subsidies-while-oil-and-gas-pro/

572 http://www.iisd.org/gsi/sites/default/files/biofuels_subsidies_eu_review.pdf

573 Pimentel, David et al. Food versus Biofuels, Human and Ecological Costs, *Human Ecology: An Interdisciplinary Journal.* Feb 2009, 37:1, 1-12.

574 http://report.mitigation2014.org/drafts/final-draft-postplenary/ipcc_wg3_ar5_final-draft_postplenary_chapter8.pdf

575 http://www.taxpayer.net/library/article/updated-taxpayer-supports-for-corn-ethanol-in-federal-legislation

576 http://ethanolproducer.com/articles/15243/eia-subsidies-for-renewables-down-as-tax-credits-diminish

577 https://news.nationalgeographic.com/2017/12/iowa-agriculture-runoff-water-pollution-environment/

578 World Resources 1994–1995: People and the Environment; New York, NY, USA, 1994.

579 https://www.eeoc.gov/laws/statutes/

580 https://www.thenation.com/article/real-story-racism-usda/

581 https://www.usatoday.com/story/news/nation/2013/03/17/women-farmers.../1993009/

582 https://www.amazon.com/Peace-Microfarms-Green-Strategy-prevent/dp/1480141208

583 https://www.amazon.com/Algae-Microfarms-community-vertical-buildings/dp/148396826X

584 Ibid.

585 https://www.ncbi.nlm.nih.gov/pmc/articles/PMC4074336/

[586] https://www.thelancet.com/commissions/EAT

[587] Emma M Brown, et al. Seaweed and human health, *Nutrition Reviews*, 72:3, March, 2014, 205–216, https://doi-org.ezproxy1.lib.asu.edu/10.1111/nure.12091

[588] Deepika, R. Charu. Phytochemical characterization of Saragassum Wightii after Phytocolloid Extraction, International Journal of Pharmaceutical, Chemical & Biological Sciences. Jul-Sep 2017, 7:3, 230-239.

[589] Ibid.

[590] www.framinghamheartstudy.org

[591] https://www.ncbi.nlm.nih.gov/pmc/articles/PMC3108955/

[592] https://www.wcrf.org/sites/default/files/Meat-Fish-and-Dairy-products.pdf

[593] https://plantbasedfoods.org/in-the-news/

[594] Iso, H. Lifestyle and cardiovascular disease in Japan, *J Atheroscler Thromb*, 2011, 18, 83 – 88.

[595] http://www.cdc.gov/reproductivehealth/maternalinfanthealth/PretermBirth.htm

[596] https://www.cdc.gov/nchs/fastats/leading-causes-of-death.htm

[597] https://www.endocrine.org/

[598] Qian Di, M.S. et al. Air Pollution and Mortality in the Medicare Population, New England Journal of Medicine, 2017; 376:2513-2522.

www.ingramcontent.com/pod-product-compliance
Lightning Source LLC
Chambersburg PA
CBHW071250220526
45468CB00001B/72